Sol-Gel Method - Recent Advances

Edited by Jitendra Pal Singh, Shakti Shankar Acharya, Sudhanshu Kumar and Shiv Kumar Dixit

Published in London, United Kingdom

Sol-Gel Method - Recent Advances
http://dx.doi.org/10.5772/intechopen.105242
Edited by Jitendra Pal Singh, Shakti Shankar Acharya, Sudhanshu Kumar and Shiv Kumar Dixit

Contributors
Srinivasa Raghavan, Maria Rosaria Plutino, Silvia Sfameni, Giulia Rando, Shrabani Mahata, Satya Sopan Mahato, Disha Mahata, Sanjibani Panda, Arafa Hassen, Adel M. El Sayed, Mohamed Shaban, Azza Al-Ghamdi, Mohamed A. Basyooni, Shrouk E. Zaki, Walid Belaid, Amina Houimi, Saptarshi Ghosh, Manoranjan Sahu, Jitendra Pal Singh, Shakti Shankar Acharya

First published in London, United Kingdom, 2023 by IntechOpen
IntechOpen is the global imprint of INTECHOPEN LIMITED, registered in England and Wales, registration number: 11086078, 5 Princes Gate Court, London, SW7 2QJ, United Kingdom

British Library Cataloguing-in-Publication Data
A catalogue record for this book is available from the British Library

Additional hard and PDF copies can be obtained from orders@intechopen.com

Sol-Gel Method - Recent Advances
Edited by Jitendra Pal Singh, Shakti Shankar Acharya, Sudhanshu Kumar and Shiv Kumar Dixit
p. cm.
Print ISBN 978-1-80355-414-3
Online ISBN 978-1-80355-415-0
eBook (PDF) ISBN 978-1-80355-416-7

Meet the editors

Dr. Jitendra Pal Singh is a Ramanujan Fellow at Manav Rachna University, Faridabad, Haryana, India. He earned his Ph.D. from the Govind Ballabh Pant University of Agriculture and Technology, Pantnagar, Uttarakhand, India. He has worked at international laboratories in India and abroad. He has published more than 150 articles in reputed journals. Dr. Singh received a Young Scientist Award from the Uttarakhand Council of Science and Technology, Uttarakhand, India, in 2009; a Young Scientist Award (Fast Track Scheme) from the Department of Science and Technology, New Delhi, India, in 2014; and a Ramanujan Fellowship Award from the Science and Engineering Research Board, New Delhi, India, in 2022. He has edited several journal issues and books and is the founder and editor-in-chief of *Prabha Materials Science Letters.*

Dr. Shakti Shankar Acharya is an assistant professor in the Department of Physics, Ravenshaw University, Cuttack, Odisha, India. Dr. Acharya obtained a Ph.D. in Physics from Siksha O Anusandhan Deemed to be University, Bhubaneswar, Odisha, India, in 2017. He worked as a post-doctoral fellow at the Institute of Physics, Bhubaneswar, India in 2017–2019. Dr. Acharya was a visiting research fellow at UGC-DAE, CSR, Indore, India, in 2013, and at Linkoping University, Germany, and the Tata Institute of Fundamental Research, Mumbai, India, in 2014. Dr. Acharya visited Julich Research Centre, Germany, in October 2022 to present his paper "Electronic Structure of Fe-Ni Invar Alloys." His research interests include magnetism and transport properties of strongly correlated systems, topological insulators, and nanomaterials.

Dr. Sudhanshu Kumar, a researcher specializing in materials science, completed his Ph.D. at the University of Delhi, India, in 2022. During that time, he conducted extensive research on the structural, electrical, and magnetic properties of magnesium ferrite (MgFe2O4) powders and ceramics. Dr. Sudhanshu demonstrated exceptional proficiency in material characterization techniques. He is an expert in the synthesis of ferrite materials via the sol–gel auto combustion technique. Dr. Sudhanshu was a postdoctoral fellow at the Technical University of Munich (TUM), Germany, in 2023–2025, under the Global Neutron Scientist (GNeuS) scheme.

Dr. Shiv Kumar Dixit is an assistant professor in the Department of Physics, Manav Rachna University, Faridabad, India. He was awarded a Ph.D. from the Department of Electronics, University of Delhi, India, in 2014. During his Ph.D. work, he was awarded a senior research fellowship from the Council of Scientific and Industrial Research. He has more than 10 years of teaching and research experience and is a Ph.D. supervisor. He has expertise in the fabrication and characterization of nanocomposite thin films. Dr. Dixit has made significant contributions to organic and hybrid solar cells, including structural, optical, and electrical properties of films and devices.

Contents

Preface

New materials and homogeneous multicomponent systems like mixed oxides can be prepared following different approaches. For example, the sol-gel method is a conventional method for synthesizing nanostructures. It can also be used to synthesize ceramics and materials used in optics, electronics, energy, surface engineering, biosensors, and pharmaceutical research.

The sol-gel process emerged in 1921 and became significant in 1960 because of its application in the nuclear industry. It allows for effortless control of the composition, properties, and architecture of nano-systems. It is a cost-effective method due to its low-temperature reaction and good control over chemical composition at the molecular level to prepare advanced materials in bulk as well as in the form of powders, fibers, and thin films. This book provides insights into the sol-gel process, with three sections on the basics and fundamentals of the method, using the method to synthesize selected materials and the various applications of these developed materials.

Jitendra Pal Singh
Department of Sciences (Physics),
Manav Rachna University,
Faridabad, India

Shakti Shankar Acharya
Department of Physics,
Ravenshaw University,
Cuttack, Odisha, India

Sudhanshu Kumar
Researcher,
University of Delhi,
Delhi, India

Shiv Kumar Dixit
Assistant Professor,
Department of Physics,
Manav Rachna University,
Faridabad, India

Section 1

Fundamentals of Sol-Gel Science

Chapter 1

Introductory Chapter: Sol-Gel Synthesis

Shakti Shankar Acharya and Jitendra Pal Singh

1. Introduction

Sol-gel technology is by far one of the best techniques for synthesis of materials for both extensive and intensive research since the inception of the materials way back in 1960 [1–5]. An improvement in the processing of conventional materials and their properties as well as novel materials synthesis are the outcomes of the sol-gel method. For the preparation of high-performance liquid chromatography, organic-inorganic hybrids' sol-gel method is very useful due to its low-temperature nature [6–8]. A schematic showing the formation of nanoparticles is depicted in **Figure 1**. Nanoparticles (magnesium ferrite ($MgFe_2O_4$)), in this case) can be transformed to bulk $MgFe_2O_4$. This reflects the ability of this method to synthesise both nanoparticles and bulk particles in a very easy process [9].

In this synthesis, metal nitrates are used as starting materials; however, sol-gel processing can also be carried out using other salts of metals (M). **Figure 2** depicts the mechanism of nanoparticle formation, either using acetate (molybdenum acetate (M(OAc)) or using halides (metal halides (MX)). Dimethyl formamide (DMF) was used as host matrix in this synthesis [10].

In brief, advantages of the sol-gel technique are as follows:

- Simple process.
- Synthesis of highly pure products.
- Synthesis efficiency is very high.
- Complex shapes of synthesis of optical components.
- Uniform composite oxides synthesis.

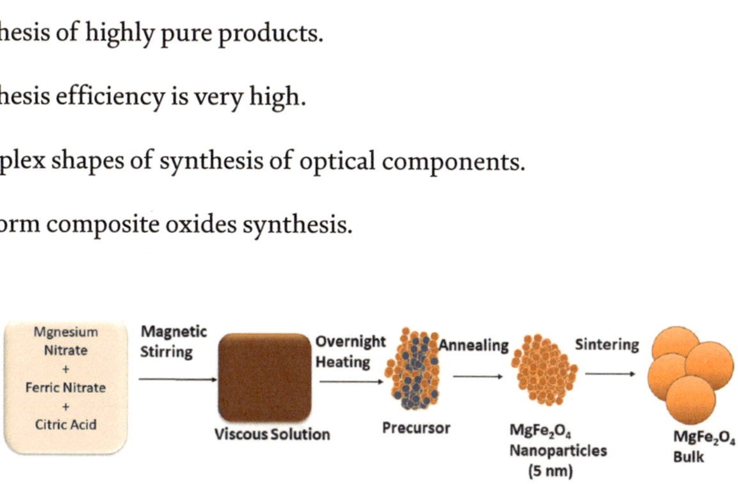

Figure 1.
Synthesis of magnesium ferrite ($MgFe_2O_4$) nanoparticles and their transformation to bulk.

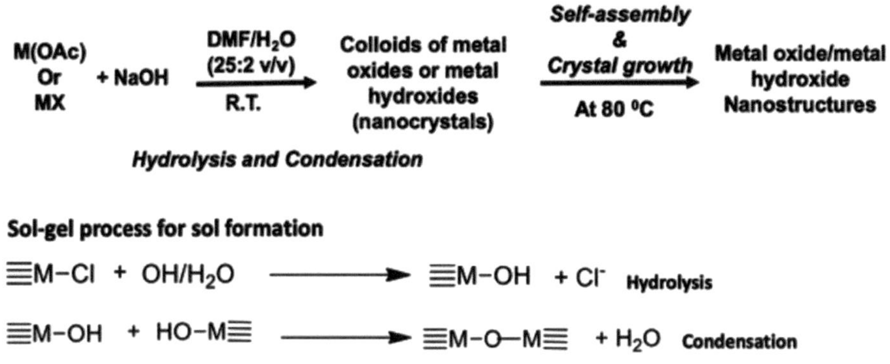

Figure 2.
Sol-gel chemical process followed by self-assembly approach for the formation of either metal oxides or metal hydroxide nanostructures. Reprinted with permission from [10].

- Tailoring of composition design and control of homogeneous materials synthesis.
- Use of specially shaped materials fibres and aerogels.
- Better surface coverage.
- Thin-layer amorphous materials synthesis.
- Low thermal expansion coefficient, low ultraviolet (UV) absorption and high optical transparency materials synthesis with tailored physical properties.
- Porous and rich materials production with organic and polymeric compounds.
- High chemical reactivity of precursors due to processing in solution phase.
- Less expensive and high-quality materials synthesis.

2. Application of sol-gel to nanoparticles

Considering these benefits, sol-gel processing is considered as a favourable processing technique for growing nanostructures. The technique is effectively used to grow metal nanoparticles [11], metal oxide nanoparticles [12, 13] and non-metallic oxide nanoparticles [14]. In the cases where certain doped elements are believed to induce impurity phases, sol-gel processing is considered very effective. For example, sol-gel processing allows higher concentration of rare earth ions inside ferrite nanoparticles with pure spinel phase [15, 16]. **Figure 3** shows the X-ray diffraction patterns (XRD) of Dy^{3+}-doped cobalt ferrite nanoparticles using sol-gel processing [17].

Some complex oxides such as lithium lanthanum zirconic oxide (LLZO), a solid state electrolyte for Li ion battery [18], can also be effectively grown using this method (**Figure 4**) [19]. This method is considered suitable to optimise phase, composition and ionic conductivity of this material [19]. Synthesis of cathodes of this battery is also reported by this method [20].

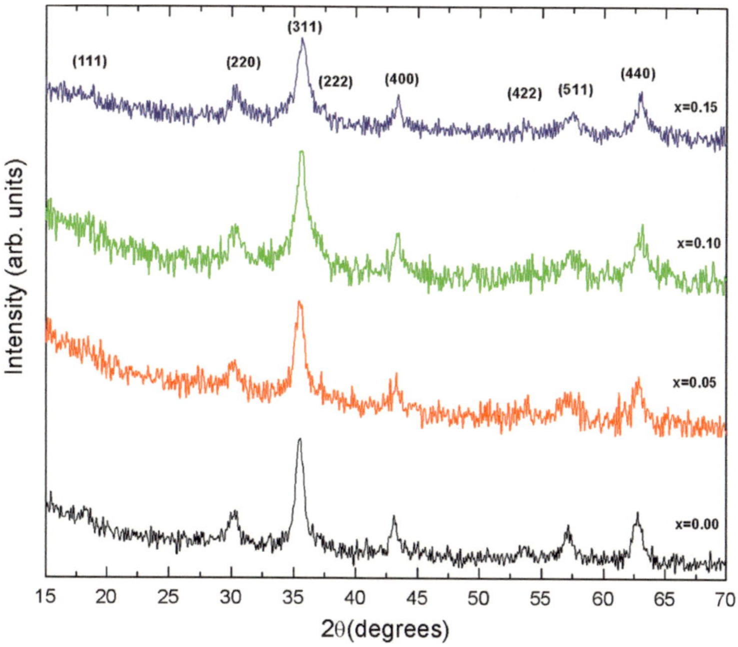

Figure 3.
Synthesis of dysprosium (Dy)-doped cobalt ferrite ($CoFe_2O_4$). No impurity phase is observed up to a Dy concentration (x) of 0.15. Reprinted with permission from [17].

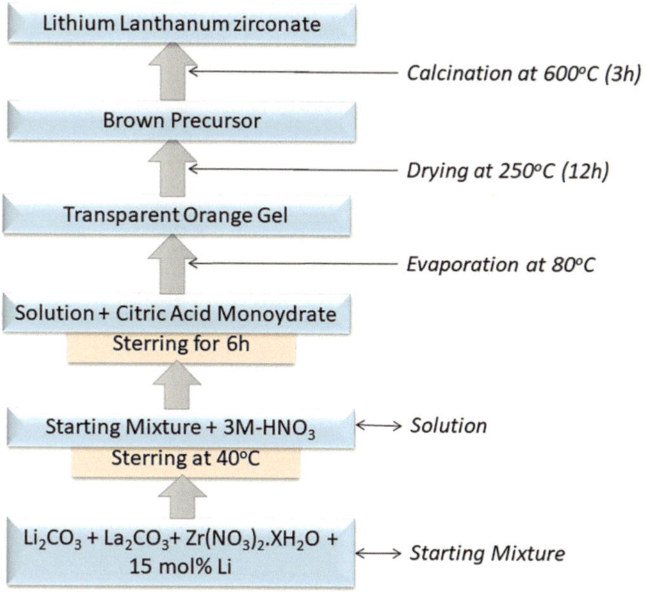

Figure 4.
Synthesis process of aluminium (Al)-doped lithium lanthanum zirconium oxide (LLZO) by taking carbonates and nitrates as starting materials. A schematic showing the formation of nanoparticles is drawn, based on the description given in the report [19].

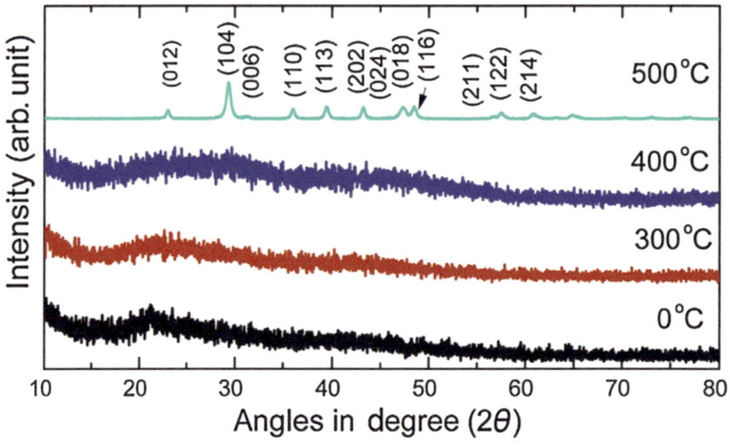

Figure 5.
Synthesis amorphous and crystalline carbonate. Reprinted with permission from [22].

Sol-gel processing is also able to produce natural minerals such as calcium carbonate in the laboratory [21, 22]. Sensitivity of annealing temperature on the phase of calcium carbonate can be seen in **Figure 5**.

This methodology is also able to produce nanostructure of different shapes, such as nanocubes [23], nanotubes [24], nanoplates [25], etc. Sol-gel processing also plays a vital role in thin film growth technology. By taking the help of dip/spin coater, thin films of appropriate material can be grown effectively [26]. Sol-gel is a low-temperature method of fabricating glass in shapes that can range from simple to very intricate. High-quality optical elements are possible, as are quality optical element-doped materials (even containing organic dyes) suitable for laser-gain media [27]. Thus, easy processing and its suitability to grow nanostructures of different materials make it favourable for commercial purposes too [28].

Acknowledgements

JPS is thankful to the Science and Engineering Board, Department of Science and Technology, New Delhi, for providing the Ramanujan Fellowship via Grant Number RJF/2021/000115.

Conflict of interest

The authors declare no conflict of interest.

Author details

Shakti Shankar Acharya[1] and Jitendra Pal Singh[2*]

1 Department of Physics, Ravenshaw University, Cuttack, Odisha, India

2 Department of Sciences (Physics), Manav Rachna University, Faridabad, India

*Address all correspondence to: jitendra_singh2029@rediffmail.com

References

[1] Abe K, Sanada Y, Morimoto T. Antireflective coatings for CRTs by sol–gel process. Journal of Sol-Gel Science and Technology. 2003;**26**:709-713

[2] Adachi T, Kawashima J, Shoshi M, Matsubara M, Sakai K, Nakasone T. Sol–gel production of silica nanoparticles. In: Schmidt H, editor. Sol–Gel Production. Z€urich: Transtech Publications; 1998. pp. 1-6

[3] Adachi T. Silica spherical microparticles applied as spacers. In: Sakka S, editor. Handbook of Solgel Science and Technology. Vol. 3. Dordrecht: Kluwer; 2004. pp. 181-189

[4] Aegerter MA, Al-Dahoudi N. Wet-chemical processing of transparent and antiglare conducting ITO coating on plastic substrates. Journal of Sol-Gel Science and Technology. 2003;**27**:81-89

[5] Agostinelli JA, Paz-Pujalt GR, Mehrotra RC. Superconducting thin films in the Bi-Sr-Ca-Cu-O system by the decomposition of metalloorganic precursors. Physica C. 1988;**156**:208-212

[6] Akamatsu Y, Makita K, Inaba H. Large size recyclable coloured glass plates prepared from organic colorant dispersed silica sols by the dipping method. Journal of Sol-Gel Science and Technology. 2000;**19**:387-391

[7] Almeida RM, Gama A, Vueva Y. Bioactive sol–gel scaffolds with dual porosity for tissue engineering. Journal of Sol-Gel Science and Technology. 2011;**57**:336-342

[8] Amberg-Schwab S. Inorganic–organic polymers with barrier properties against water vapor, oxygen and migrating monomers. In: Sakka S, editor. Handbook of Sol–Gel Science and Technology. Vol. 3. Dordrecht: Kluwer; 2004. pp. 455-478

[9] Singh JP, Kim SH, Won SO, Lim WC, Lee IJ, Chae KH. Covalency, hybridization and valence state effects in nano-and micro-sized $ZnFe_2O_4$. CrystEngComm. 2016;**18**:2701-2711

[10] Yarbrough R, Davis K, Dawood S, Rathnayake H. A sol–gel synthesis to prepare size and shape-controlled mesoporous nanostructures of binary (II–VI) metal oxides. RSC Advances. 2020;**10**:14134-14146

[11] Gao H, Xiang W, Ma X, Ma L, Huang Y, Ni H, et al. Sol-gel synthesis and third-order optical nonlinearity of Au nanoparticles doped monolithic glass. Gold Bulletin. 2015;**48**:153-159

[12] Parashar M, Shukla VK, Singh R. Metal oxides nanoparticles via sol–gel method: A review on synthesis, characterization and applications. Journal of Materials Science: Materials in Electronics. 2020;**31**:3729-3749

[13] Niederberger M. Nonaqueous Sol–Gel Routes to Metal Oxide Nanoparticles. Accounts of Chemical Research. 2007;**40**(9):793-800

[14] Huang T, Zhang L, Chen H, Gao C. Sol–gel fabrication of a non-laminated graphene oxide membrane for oil/water separation. Journal of Materials Chemistry A. 2015;**3**:19517-19524

[15] Kumar H, Negi P, Singh JP, Srivastava RC, Ambreen S, Asokan K. Tuning of structural, electrical and transport behaviour of cobalt nanoferrite by dysprosium ions substitution. Ceramics International. 2023;**49**(16):27294-27302

[16] Dixit G, Singh JP, Srivastava RC, Agrawal HM. Magnetic resonance study of Ce and Gd doped $NiFe_2O_4$ nanoparticles. Journal of Magnetism and Magnetic Materials. 2012;**324**:479-483

[17] Kumar H, Srivastava RC, Pal SJ, Negi P, Agrawal HM, Das D, et al. Structural and magnetic study of dysprosium substituted cobalt ferrite nanoparticles. Journal of Magnetism and Magnetic Materials. 2016;**401**:16-21

[18] Singh JP. Materials towards the development of Li rechargeable thin film battery. Prabha Materials Science Letters. 2023;**2**:26-40

[19] Košir J, Mousavihashemi S, Wilson BP, Rautama E-L, Tanja K. Comparative analysis on the thermal, structural, and electrochemical properties of Al-doped Li7La3Zr2O12 solid electrolytes through solid state and sol-gel routes. Solid State Ionics. 2022;**380**:115943

[20] Etacheri V. Sol-gel processed cathode materials for lithium-ion batteries. In: Pillai S, Hehir S, editors. Sol-Gel Materials for Energy, Environment and Electronic Applications. Advances in sol-Gel Derived Materials and Technologies. Cham: Springer; 2017. DOI: 10.1007/978-3-319-50144-4_6

[21] Singh V, Paidi AK, Shim C-H, Kim S-H, Won S-O, Singh JP, et al. Calcite nanocrystals investigated using X-ray absorption spectroscopy. Crystals. 2012;**11**(5):490

[22] Singh JP, Ji M-J, Shim C-H, Kim SO, Chae KH. Effect of precursor thermal history on the formation of amorphous and crystalline calcium carbonate. Particulogy. 2017;**33**:29-34

[23] Kaifeng Y, Guo Y, Ding X, Zhao J, Wang Z. Synthesis of silica nanocubes by sol–gel method. Materials Letters. 2005;**59**(29-30):4013-4015

[24] Anastasescu C, Mihaiu S, Preda S, Zaharescu M. Synthesis of oxide nanotubes by sol–gel method. In: 1D Oxide Nanostructures Obtained by sol-Gel and Hydrothermal Methods. Cham: Springer Briefs in Materials. Springer; 2016. DOI: 10.1007/978-3-319-32988-8_2

[25] Liu Y, Lv H, Jiayuan H, Li Z. Synthesis and characterization of Bi_2WO_6 nanoplates using egg white as a biotemplate through sol-gel method. Materials Letters. 2015;**139**:401-404

[26] Plenet JC et al. Preparation of thin films using a sol-gel method. In: Caliste JP, Truyol A, Westbrook JH, editors. Thermodynamic Modeling and Materials Data Engineering. Data and Knowledge in a Changing World. Berlin, Heidelberg: Springer; 1998. DOI: 10.1007/978-3-642-72207-3_27

[27] Mukherjee SP. Sol-gel processes in glass science and technology. Journal of Non-Crystalline Solids. 1980;**42**(1-3):477-488

[28] Hong YJ, Yi GR. Industrial applications of sol-gel technology. SSP. 2007;**124-126**:619-622. DOI: 10.4028/www.scientific.net/ssp.124-126.619

Chapter 2

Perspective Chapter: Sol-Gel Science and Technology in Context of Nanomaterials – Recent Advances

Satya Sopan Mahato, Disha Mahata, Sanjibani Panda and Shrabani Mahata

Abstract

Sol-gel method is a novel technology of producing new materials in a convenient and cost-effective way. This method allows a highly ordered and well-connected network structure to be developed and better controlled. It is a simple procedure to produce homogenous multi-component systems. Homogenous mixed oxides can be developed by combining different molecular precursor solutions. The advantages of sol-gel method include its simplicity, affordability, controllability, and ability to mass production of nano-sized particles with large surface areas. Due to this simplicity and versatility, sol-gel technology has higher admiration and industrial application compared to many prevailing methods and is widely being used in various fields. Sol-gel procedure has been comprehensively used as a common and practical way for the development of nano-structured materials for a wide range of applications. This chapter primarily concentrates on the fundamentals of sol-gel science, particularly with respect to the development of nanoparticles, and their numerous applications, with a focus on more recent, sophisticated, and advanced applications.

Keywords: *sol-gel science*, novel technology, *hydrolysis*, *condensation*, *network structure*

1. Introduction

The sol-gel method initially appeared in the realm of nanoscience and technology in the year 1921. Its development happened in the 1960s as a consequence of the nuclear industry's obligation to new synthesis techniques. This development gained admiration around 1984 and had substantial development up to the year 2011. An essential guide for the development of the sol-gel method was provided by Dr. Jeffrey Brinker, a pioneer in the synthesis of materials and sol-gel science [1]. Dongyuan Zhao and David Avnir were two distinguished researchers who have advanced their studies in the field of sol-gel science and made noteworthy contributions in this field [2]. The process of sol-gel synthesis is so widely [3–11] used that it has become the most attractive technology for versatile applications.

The sol-gel chemistry is based on poly-condensation and hydrolysis processes. Due to their propensity to create homogenous solutions in a wide range of solvents, as well as due to their reactivity toward nucleophilic reagents, such as water and metal alkoxides [$M(OR)_3$], it is a useful process to generate oxides. Metal oxides, nitrides, and carbides are just a few examples of ceramic materials that are commonly prepared using sol-gel techniques [12–17]. This method has a number of benefits over traditional processing technologies [18–21], including low reaction temperatures [5, 22–27], precise composition control [28, 29], high levels of purity [30–32], and the capacity to create processes for large-area applications [33–41]. The key steps of the reaction are the hydrolysis and condensation of an inorganic or metal-organic precursor, which produces a sol that, following a series of chemical reactions and/or mild thermal treatments, transforms into a gel, which is then calcined to produce the end product.

The final product and its applications were the main goals of the majority of studies that were reported, with little regard for the conditions of synthesis or the reaction mechanisms used to produce gels, even though the characteristics of a gel and how it responds to heat treatment may be highly dependent on the structure that was already established at the sol stage [42–46]. The primary characteristics of the resultant powder are thus determined by the generation of colloidal aggregates. The structure and shape of samples generated are significantly changed by altering the chemical conditions under which materials are polymerized. A few applications of sol-gel-based nanomaterials are shown in **Figure 1**.

Sols are thermodynamically unstable, due to their large surface area to volume ratio. However, they can be stabilized by introducing an energy barrier on the solid particles present in the sol. This stabilization process can be done either by electrostatic stabilization or steric stabilization.

Although it was researched earlier, sol-gel chemistry has been the subject of substantial research since the mid-1970s, when it was discovered that metal alkoxide solutions can be used as starting materials for sol-gel processes that build a variety of inorganic networks [33–36]. Sol-gel processing enables the production of

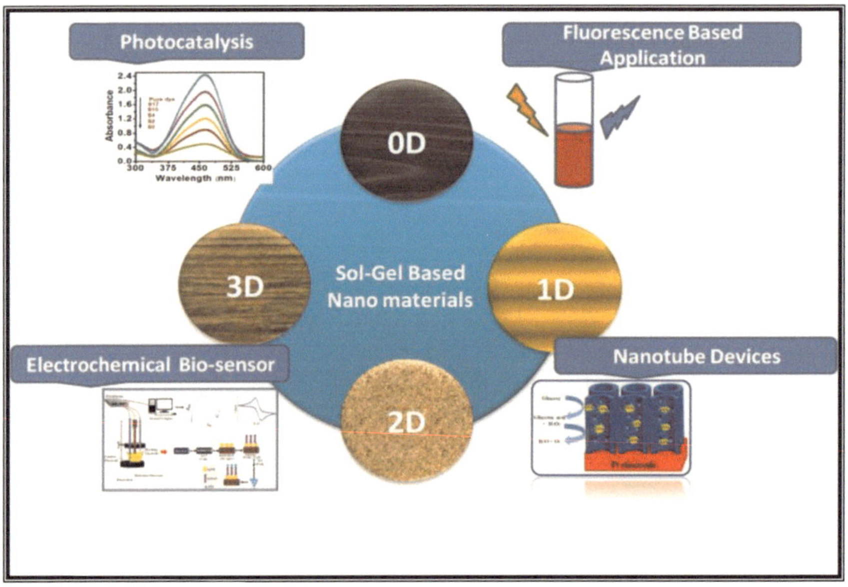

Figure 1.
Applications of sol-gel-based nanomaterials.

homogeneous, high-purity inorganic oxide glasses at room temperature as opposed to the extremely high temperatures necessary for traditional glass creation. Products have been developed for use in a variety of fields, including gas separations, elastomers, coatings, and laminates [34, 37–42]. These products include molded gels, spun fibers, thin films, molecular cages, and xerogels. Numerous property adjustments can be made by incorporating inorganic compounds into organic polymers.

Sol-gel-based nanomaterials have been considered as being of high technological impotence compared to many other materials [43–46]. In order to have strong control over the properties of the finished material, the goal of this effort is to better understand the chemistry involved in the creation of nanomaterials *via* the sol-gel technique. This chapter largely focuses on the principles of sol-gel science with an emphasis on more modern, sophisticated, and advanced applications, notably with regard to the creation of nanoparticles, thin film, fibers, and its many advanced applications.

2. Importance of sol-gel technology

Years after its discovery, sol-gel science and technology still continue to draw attraction of researchers from around the globe. Sol-gel method is a bottom-up synthesis technique. Sol-gel surface modification of substrates is very successful and in high demand [47–54]. The main benefit of the sol-gel process is that it produces stable surfaces with a high surface area. A novel method of creating new materials is offered by the sol-gel technique. This approach differs from others in a few key ways. Low processing temperature, great homogeneity, and simplicity characterize this process. Its compatibility with polymers and polymerization allows it to produce nanoparticles in an organic environment and is one of its key advantages. This process results in high-purity products. Through this method, the entire sequence of processes involved in the synthesis of solids can be better controlled. It is simple to create homogenous multi-component systems; in particular, homogenously mixed oxides can be created by combining molecular precursor solutions. Although all of the technologies can produce large amounts of nanomaterial, the sol-gel process is more widespread and has more industrial uses than other existing technologies [55–60]. This technology can create high-quality identical nanoparticles in commercial quantities due to their unique traits and properties [61–65]. Functional materials like photocatalysts [66–71], ferroelectrics [72–79], nonlinear optical materials [80–84], and superconductors [85–93] can be made using the sol-gel process. Therefore, significant contribution is continuing throughout the world still today. Well-cited articles on sol-gel science and technology published in 2020 (Source: SCOPUS) are shown in **Figure 2**.

Metal NPs-doped inorganic-organic hybrid films, such as organically modified silica, are created using the sol-gel method. Metal oxides, nitrides, and carbides are just a few examples of ceramic materials that are commonly prepared using sol-gel techniques. The sol-gel method can be used to create ceramics as a molding substance and as a transitional layer between thin metal oxide coatings in a variety of applications. Numerous optical, electrical, energy, surface engineering, biosensor, medicinal, and separation technologies utilize the materials produced by the sol-gel process [55, 94–98]. To connect researchers in the field of science and technology all over the world International Sol-Gel Society (ISGS) has been formed. It is an international, interdisciplinary, not-for-profit organization whose primary purpose and objective are the advancement of sol-gel science and technology. International Sol-Gel Society member/subscriber affiliation in 2020 is shown in **Figure 3** [97].

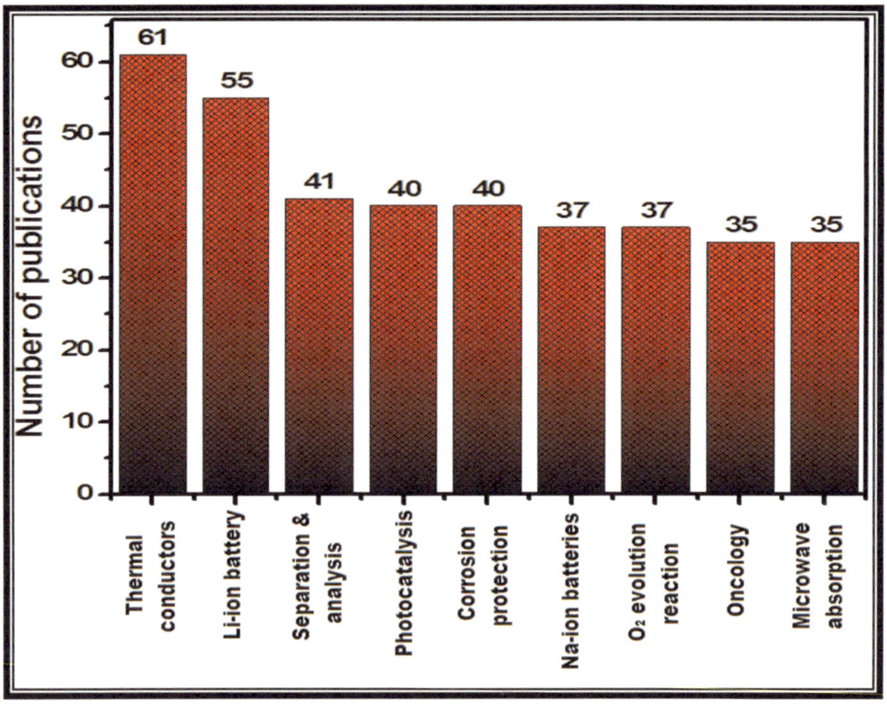

Figure 2.
Well-cited articles on sol-gel science and technology published in 2020 (source: SCOPUS).

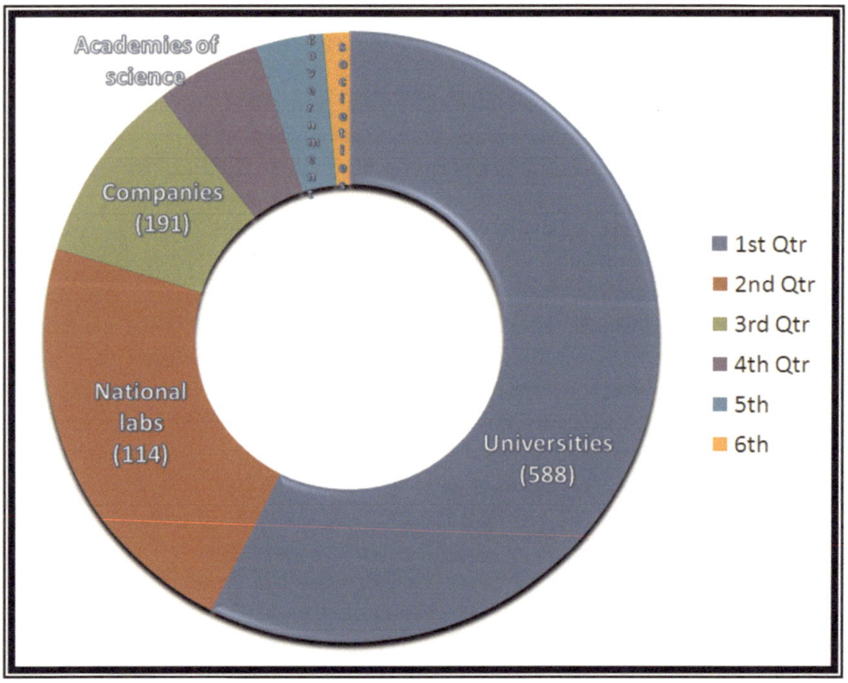

Figure 3.
International sol-gel society member/subscriber affiliation in 2020 [97].

Sol-gel method has a number of benefits over traditional processing technologies, including low reaction temperatures, precise composition control, high levels of purity, and the capacity to create processes for large-area applications. It is a well-known technique for a variety of materials, including inorganic membranes, monolithic glasses and ceramics, thin films, and ultra-fine powders. Even for the synthesis of 1D nanomaterials the process is used today. The numbers of merits of this method have made it highly attractive, which can be carried out at room temperature with the potential for simple chemical doping. The aforementioned examples show that sol-gel methods are simple processes to modify the final morphological characteristics of nanoparticles. Alloy products can be made in a single step by combining two or more metal (or metal oxide) precursors in a precise ratio since this technique can simultaneously make two or more different types of nanoparticles. There are alternative single-step procedures that can produce alloy products, such as the plasma method and electrochemical processes, but the key feature that sets the sol-gel method apart from them is its industrial scale. The sol-gel process can also be used to produce highly homogenous composites with a 99.99% purity level. Metal and ceramic nanoparticles can be produced with this technology at temperatures between 70 and 320°C because the process temperature is lower than with current methods. Different ways of sol-gel synthesis are shown in **Figure 4**.

The main advantages of the sol-gel method include superior product purity, narrow particle size distribution, and production of a homogenous nanostructure at low temperatures. Metal nano-oxides are frequently created using this technique [29, 99–101]. The sol-gel process, as mentioned, entails changing a sol into a gel using a variety of methods, the majority of which call for mild drying to get rid of the solvent. The ability to prevent cracks from forming is one of the most important properties of this technology. It can produce integrated components by molding and curing the gel that is produced. Products made by molding or casting are utilized as filters and membranes.

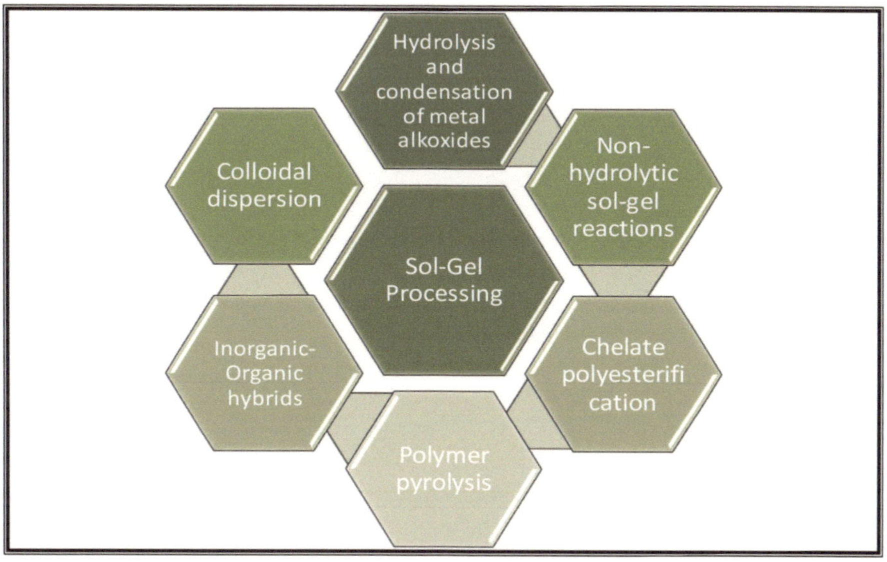

Figure 4.
Different ways of sol-gel synthesis.

Thin films with a thickness of 50–500 nm can be created using the sol-gel method. Sol-gel thin films are used in both the chemical and electronic sectors for a range of applications. Coatings created using the sol-gel technique also affect the material's optical properties. Materials that are composite or nanocomposite can be created using the sol-gel technique. This is accomplished by loading secondary components into continuous porosity at the nanoscale. To create denser parts, the produced or synthesized pieces are put through sintering operations. Because the large specific surface area accelerates the rate of compaction or compaction of the structure, the nanoporous gels compact better and more quickly during the sintering process. But it is crucial to remember that increasing the temperature during this process encourages grain growth and results in a microstructure with coarse grains.

3. What is sol and gel?

3.1 What is sol?

In a liquid phase, a sol is a stable suspension of colloidal solid particles. These solid particles are small enough and denser than the liquid they are in, so the forces of dispersion outweigh the forces of gravity. Lyophilic and lyophobic sols are subclassified. A lyophobic sol is one in which the solvent-particle interaction is relatively weak, whereas a lyophilic sol is one in which this interaction is quite strong. When the forces between two particles are repellent and the surface's assisting forces keep the particles from agglomerating and coagulating, sols become stable. When another additive removes the particle's charge, flocculation occurs and the system disintegrates, which causes gel to develop.

3.2 What is sol-gel?

A gel is a colloidal dispersion of tiny particles in a liquid. It is a translucent, porous, -three-dimensional solid network that is interconnected and porous. In other terms, gel is a two-part, liquid system with a semi-solid character. A sol or solution turns into a gel during the process of gelation, and the gel's continuous solid structure provides it flexibility. These solid particles that are present in the gel can be macromolecules, amorphous solids, or crystalline solids. The gel is recognized as colloidal when the solid network is composed of colloidal sol particles and as polymeric when the solid network is composed of sub-colloidal chemical units. Colloidal gels and polymeric gels differ primarily in that the former's sol-gel transition is brought on by a physicochemical effect, while the latter is brought on by chemical bonding as opposed to a chemical reaction like poly-condensation. Sols have a high surface area-to-volume ratio, which makes them thermodynamically unstable.

4. Sol-gel processes

For the development of various types of nanostructures, particularly metal oxide nanoparticles, sol-gel method is one of the most attractive and widely used wet chemical methods. The hydrolysis of a precursor-containing solution yields suspended colloidal particles, which is the basis of this procedure. The creation of a homogenous sol from the precursors and its transformation into a gel is the foundation of the sol-gel process.

In the conventional procedure of sol-gel procedure, the molecular precursor mainly a metal alkoxide is dissolved in water or alcohol and stirred until it gels. Since the gel created during the hydrolysis process is wet in nature, it needs to be dried appropriately depending on its use and desired qualities. For instance, burning alcohol is used to complete the drying process if the solution is alcoholic. The generated gels are pulverized and then calcined after the drying stage. Due to the low reaction temperature and cost-effectiveness of the sol-gel process, the chemical composition may be controlled well. The leftover gel is then dried after the solvent in the gel is removed from the gel structure. The dried gel's characteristics are highly dependent on the drying method. In other words, the "removing solvent method" is chosen in line with the intended use of the gel. Industries including surface coating, building insulation, and the creation of specialty garments all use dried gels in various ways. It is important to note that nanoparticles can be produced by crushing gel in specialized mills.

- A stable colloidal solution termed sol is first produced during the sol-gel process.
- The sol is a liquid suspension of 1 nm to 1 micron-sized solid particles.
- It can be made by partial condensation and hydrolyzation of precursors of an inorganic salt or a metal alkoxide.
- A gel is developed when sol particles are further condensed into a three-dimensional network.
- In the gel, which is a diphasic substance, the solids enclose the solvent.
- The generated oxide species have a steadily rising molecular weight. When using water as a solvent, the substances are known as an aqua sol or aqua gels, and when using alcohol, they are known as aquosol or alcogel.
- The product that results from evaporating gels is known as xerogel.
- Aerogels are the dried gels produced by supercritical drying of gels. High porosity and high pore volume are retained by the aerogel.

Different forms of gel formation are shown in **Figure 5**.

The sol-gel method is distinguished from other routes of material preparation from solutions or melts such as precipitation and crystallization by two main characteristics:

- Formation of clear colloidal solution due to primary condensation of dissolved molecular precursors.
- These colloidal particles merge during subsequent gelation stage into polymeric chain by chemical bonding between local reactive groups at their surface.

Formation of a metal oxide involves connecting the metal centers with oxo (M-O-M) or hydroxo (M-OH-M) bridges, therefore generating metal-oxo or metal-hydroxo polymers in solution.

The sol-gel process allows synthesizing ceramic materials of high purity and homogeneity by a process that occurs in a liquid solution of organometallic precursors

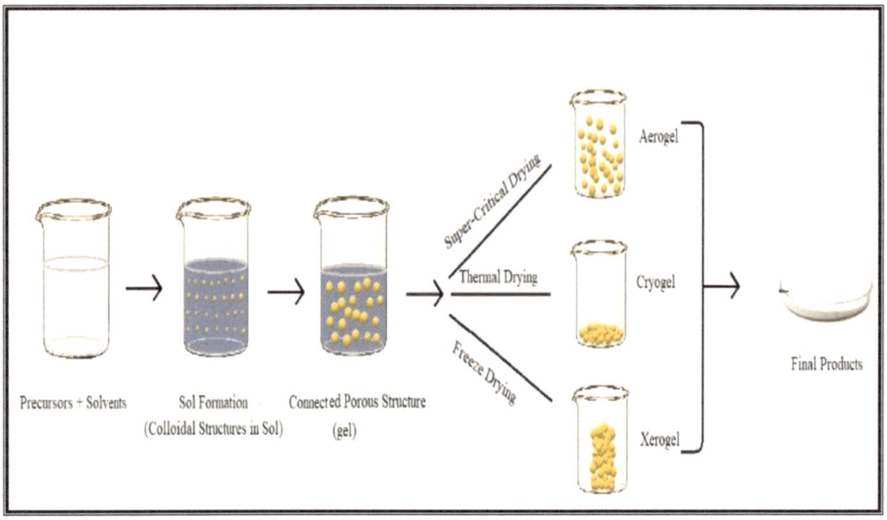

Figure 5.
Different forms of gel formation through the sol-gel method.

(TMOS, TEOS, Zr (IV)—Propoxide, Ti(IV)—Butoxide, etc.), which, by means of hydrolysis and condensation reactions, lead to the formation of a new phase (SOL).

$$\text{Step 1: (Hydrolysis) } M{-}O{-}R + H_2O \rightarrow M{-}OH + R{-}OH \; (M = Si, Zr, Ti). \quad (1)$$

$$\text{Step 2: (Water Condensation) } M{-}OH + HO{-}M \rightarrow M{-}O{-}M + H_2O. \quad (2)$$

$$\text{Step 3: (Alcohol Condensation) } M{-}O{-}R + HO{-}M \rightarrow M{-}O{-}M + R{-}OH. \quad (3)$$

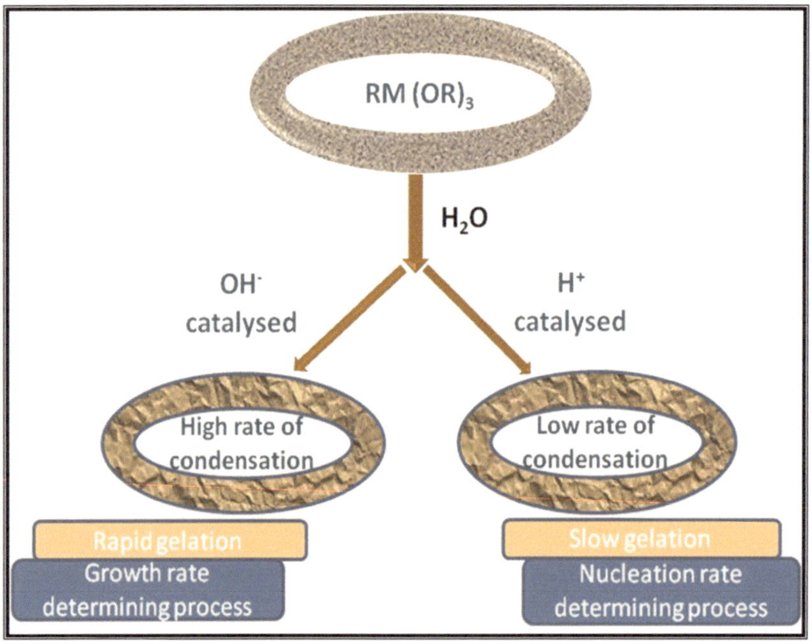

Figure 6.
Effect of catalyst on hydrolysis and condensation in sol-gel method.

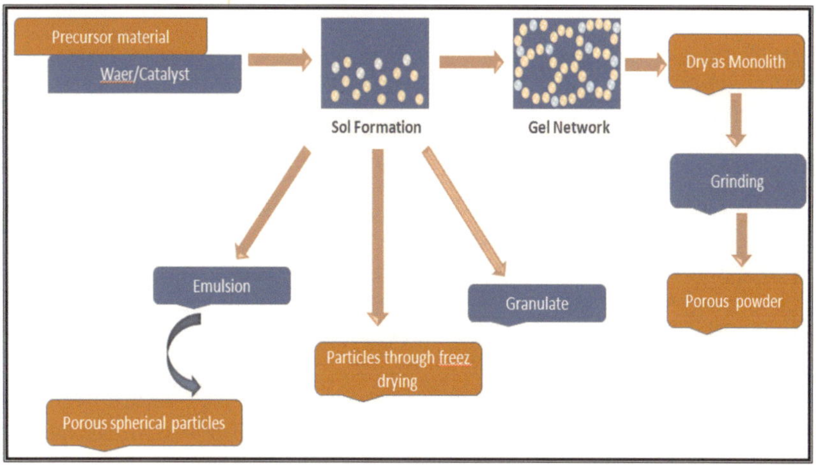

Figure 7.
Different forms of materials formation through sol-gel method.

Effect of catalyst on hydrolysis and condensation in sol-gel method is shown in **Figure 6**.

A summary of the sol-gel method's steps is shown in **Figure 7**.

4.1 Hydrolysis step

The sol-gel method is a chemical process used to produce solid materials from small molecules, typically metal alkoxides. The hydrolysis step of the sol-gel process involves the reaction of a metal alkoxide with water to produce metal hydroxide and alcohol.

The hydrolysis reaction occurs in the presence of a catalyst, typically an acid or base, which promotes the reaction between the metal alkoxide and water. The reaction can be represented by the following equation:

$$M(OR)n + nH_2O \rightarrow M(OH)n + nROH. \quad (4)$$

Where M represents a metal atom, R represents an organic group, and n is the number of organic groups attached to the metal atom.

During the hydrolysis step, the metal hydroxide and alcohol formed can further react to form a gel-like network of metal oxide particles. This gel network can then be dried and calcined to produce a solid material.

The hydrolysis step is crucial in the sol-gel process as it determines the size and morphology of the resulting metal oxide particles and the properties of the final product. Control of the reaction conditions, such as the pH and temperature, is important to achieve the desired product properties.

4.2 Condensation step

The condensation step of the sol-gel method involves the reaction of metal hydroxide species formed during the hydrolysis step to form metal oxide particles. The condensation reaction occurs in the presence of a catalyst or under controlled conditions such as temperature, pH, and solvent choice.

The condensation reaction can be represented by the following equation:

$$M(OH)_n \rightarrow MO_{n/2} + (n/2)H_2O. \quad (5)$$

Where M represents a metal atom and n is the number of hydroxyl groups attached to the metal atom.

The reaction involves the elimination of water molecules and the formation of metal-oxygen-metal bonds, leading to the formation of a network of metal oxide particles. The resulting structure can range from discrete particles to a continuous gel network depending on the reaction conditions.

The properties of the resulting metal oxide particles are influenced by the reaction conditions during the condensation step. Factors such as temperature, pH, and catalyst concentration can affect the particle size, morphology, and crystallinity of the final product.

The condensation step is a critical step in the sol-gel process as it determines the structure and properties of the final product. Careful control of the reaction conditions is necessary to produce materials with desired properties.

4.3 Olation step

The olation step in the sol-gel method involves the reaction of metal-organic complexes or metal carboxylates with a base to form a stable metal-oxide precursor. This step is also referred to as deprotonation, and it is typically used to improve the reactivity and stability of the precursor.

The reaction can be represented by the following equation:

$$M(RCOO)_n + 2NaOH \rightarrow M(OH)_n + nRCOONa + H_2O \quad (6)$$

Where M represents a metal atom, R represents an organic group, and n is the number of organic groups attached to the metal atom.

The reaction occurs in the presence of a solvent, typically water or alcohol, and a catalyst such as a base. The metal-oxide precursor formed during the olation step is often more reactive and stable than the metal-organic complex and can be used to produce a sol.

The olation step is critical in the sol-gel method as it helps to improve the reactivity and stability of the precursor, which in turn affects the properties of the final product. The properties of the metal-oxide precursor are dependent on the nature of the metal precursor, the type of organic acid used, and the reaction conditions such as temperature and pH.

Overall, the sol-gel method is a versatile approach for synthesizing metal oxide materials with unique properties, and the olation step is an essential component of this process.

4.4 Oxolation step

The oxolation step in the sol-gel method involves the reaction of metal alkoxide or metal salt with organic acid or carboxylic acid to form a stable metal-organic complex. This step is also known as chelation, and it is typically used to improve the stability of the sol by preventing the hydrolysis of the metal precursor.

The reaction can be represented by the following equation:

$$M(OR)_n + nRCOOH \rightarrow M(RCOO)_n + nROH \quad (7)$$

Where M represents a metal atom, R represents an organic group, and n is the number of organic groups attached to the metal atom.

The reaction occurs in the presence of a solvent, typically alcohol, and a catalyst such as a base. The metal-organic complex formed during the oxolation step is soluble in the solvent and can be used as a precursor to form a sol. The oxolation step is often followed by the hydrolysis and condensation steps to produce a gel-like network of metal oxide particles.

The oxolation step is critical in the sol-gel process as it helps to improve the stability of the sol, which in turn affects the properties of the final product. The stability of the sol is dependent on the nature of the metal precursor, the type of organic acid used, and the reaction conditions such as temperature and pH.

Overall, the sol-gel method is a versatile approach for synthesizing metal oxide materials with unique properties, and the oxolation step is an essential component of this process.

4.5 Sol formation

The hydrolysis reaction forms the foundation of this phase. A small amount of water is introduced to the reaction medium to start the hydrolysis reaction of precursor material. Large molecules are broken down into smaller parts by the consumption of water in the hydrolysis reaction. When a homogenous solution is developed in a water-free solvent, water is added. When water is present, the precursor undergoes a hydrolysis reaction that somehow activates it, causing the metal oxide particles to form into small, solid particles that are disseminated in the solvent resulting in the formation of sol. In a true sol, the solute is equally dispersed as an atom, molecule, or ion in the solvent, and the particle size is less than 1 nm. When the size of the particles ranges between 1 and 100 nanometers, the mixture is referred to as a colloid, and they typically remain dispersed. The sol is actually a solution or, more correctly, a colloidal mixture since it is made up of extremely small particles (less than 100 nm) scattered in the solvent phase.

4.6 Gel formation

It just takes a small amount of stimulation to the solution to cause the distributed tiny particles, each of which contains many molecular or atomic units of the respective precursors, to start forming a gel [66]. A suitable reagent can be used to perform this stimulation (pure water or water with NaCl and NaOH). Units of tens of thousands of molecules line up together to form an infinitely large, three-dimensional molecule that fills the entire volume of the reaction vessel by causing physical and chemical interactions between suspended particles and the solution. Wet gel is a massive molecule that holds all the solvent within its numerous holes. Sol-to-gel conversion is an inorganic polymerization reaction whose ultimate product is an oxide network including MOM metal oxide clusters, controlled by reactions referred to as condensation. When two simple molecules join to create a more complex molecule in the compaction reaction, a tiny molecule, like water, is released developing a long-range well-ordered three-dimensional network structure. Depending on the solvent employed or the drying technique, several types of gels are formed, and they have distinct properties and uses. Colloidal network formation in sol-gel materials is shown in **Figure 8**.

4.7 Electrostatic stabilization of the sol

The electrostatic repulsion caused by charges adsorbed on the particles and attractive Van der Waals forces are combined to provide the net force between the particles in a sol. In a liquid medium, two microscopic particles develop a Van der Waals interaction. Some particular ions are frequently preferentially adsorbed on the surface of solid particles when disseminated in a liquid medium that also contains an electrolyte. Near the charged surface, a surface charge forms in the solution. These dispersed layers that overlap and encircle two sol particles repel one another. If the van der Waals force of attraction is greater than the repelling forces caused by the surface charges, the sol will be stable. The two types of ions that make up the double layer—charge-determining ions, which regulate the charge on the surface of the particles, and counter ions, which are present in the solution close to the particle—are what create the repulsive barrier. The potential difference between the two layers is known as "zeta potential," and the counter ions serve as screen charges for the determining ions' potential. The size of the zeta potential is also connected to a colloid's stability. The point of zero charge is the pH at which a particle is neutrally charged, and the isoelectric point is the potential at which the value of zeta potential is zero (PZC).

4.8 Electro steric stabilization of the sol

By adhering sterically packed organic molecules to the surface of the colloidal particles, steric stabilization can be accomplished. This organic coating that has been adsorbed creates a steric barrier that prevents the two particles from coming too close to one another in terms of both enthalpy and entropy. The adsorbed layer must fulfill the following criteria in order to function as a barrier effectively:

1. The particle's surface must be entirely covered.

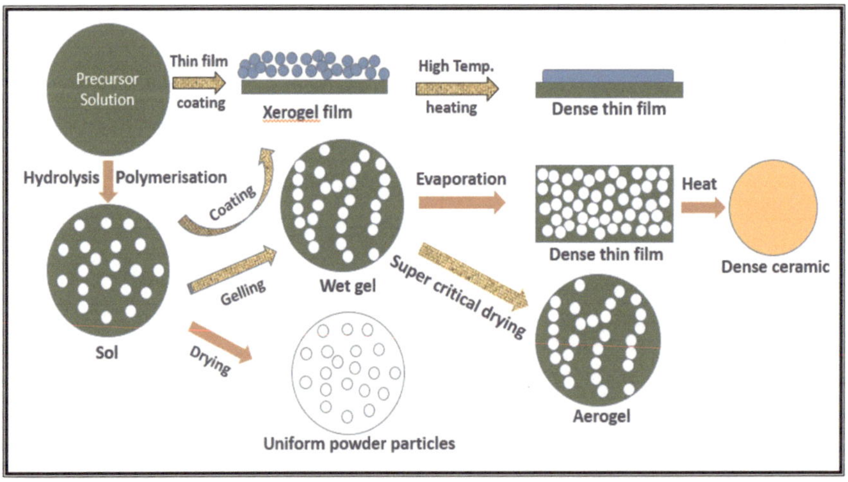

Figure 8.
Colloidal network formation in sol-gel materials.

2. The organic molecules should be chosen so that the resulting energy barrier is greater than or substantial enough to defeat the alluring van der Waals forces.

Stabilization by electrostatic effects can result from the combination of steric stabilization and electrostatic stabilization.

5. Factors influencing the sol-gel reaction

The hydrolysis and condensation reactions can be affected by a variety of factors. The water-to-alkoxide ratio, the kind and quantity of the catalyst, the kind of organic group(s), pH, and solvent effects are among those that are most important. These elements will now be taken into account.

5.1 Water-to-alkoxide ratio

The water-to-alkoxide ratio has an impact on the sol-gel process such that as the ratio rises, the amount of oxide formation increases. As a result, each alkoxide group requires at least one mole of water for full hydrolysis. Some studies have gone even further, asserting that re-esterification, which is the reverse process, will happen more quickly than the forward reaction if more than one mole of water is utilized for each alkoxide group. The water/alkoxide ratio did not correlate with the accomplishment of complete hydrolysis in a recent publication by Mc Cormick et al. Because water is produced in the process *in situ* and the reaction, once catalyzed, self-propagates the hydrolysis, it is believed that these investigations on the influence of the water/alkoxide ratio are true and accurate.

5.2 Type of network modifier

Organically modified oxides are produced when the precursor is combined with a capping agent that is chemically linked directly to the metal atom. Co-ordination centers are caused by organic groups and condensation functionality of less than four and affect the connectivity of the sol-gel network in two ways by altering the reactivity of the alkoxy groups. The diffusion of partly hydrolyzed molecules is necessary for the formation of oxide linkages. The rate of diffusion is slowed down and, as a result, there is less interconnection within the network. It is also claimed that polymers with greater surface areas are produced by alkyl groups with a larger size through this network formation. Higher concentrations of unreacted oxide are possible in gels with larger surface areas. The sol-gel network experiences a branching effect as a result. Along with these two aspects, it has also been previously mentioned that steric factors, such as the presence of bulky and/or lengthy alkyl substituents, may slow down the condensation process impacting the network formation.

5.3 Solvent effect

The regulation of hydrolysis and condensation rates can be significantly impacted by the addition of exogenous solvents. Tetrahydrofuran, formamide, dimethylformamide, and oxalic acid are among the solvents that fall under the category of drying chemical control additives, which exclude the typical cosolvents of water and alcohol. Solvent type may have a significant impact on condensation rates. Protic solvents appear to have little to no impact on the rate of condensation. By placing the negative end of the polar aprotic around the cation and adding unshared electron pairs to the cation's empty orbitals, the cation is solvated. The transition state is stabilized by this mechanism, which also speeds up condensation. For sol-gel reactions inside a membrane template, the use of DCCAs may be advantageous.

5.4 Type and amount of catalysis

The presence of a catalyst can speed up a chemical reaction. This is extremely pH sensitive in a lot of sol-gel chemistry. This is due to the fact that whereas bases (OH) and acids (H^+) are both catalysts, they do so through various processes. Whether a specific concentration is required should be taken into account when choosing the catalyst concentration. Water is produced *in situ* during the sol-gel process through condensation processes. This makes it challenging to calculate and add any precise amount of catalyst. McCormick et al. [97] generated sol-gel films with a variety of acid concentrations. According to their findings, there is no connection between acid content and the start of the sol-gel process. On-site production of water reduced the initial acid content in each trial. It was found from these experiments that only a catalytic amount of acid was required. Therefore, the reaction may self-propagate if there was this little catalyst present in all experiments. Although the entire network's fundamental structure would not be altered, the kinetics of the reaction may shift as a result. The majority of inorganic alkoxides may undergo hydrolysis and condensation reactions without the assistance of a catalyst due to their incredibly high reaction speeds. On the other hand, some alkoxides hydrolyze considerably more slowly, necessitating the inclusion of either an acid or a base catalyst. Due to the quick hydrolysis, particle nucleation rate-determining processes using acid catalysts tend to produce more linear-like networks. Contrarily, base-catalyzed reactions result in very dense materials because the sol particles have more time to assemble and arrange themselves in the most thermodynamically stable configuration.

5.5 pH

pH has a significant impact on any colloidal chemistry involving water. As sol-gel chemistry is catalyzed by the presence of acid, the presence of H^+ catalyzes comprehensively affects hydrolysis and condensation reaction rates and sol-gel chemistry is extremely pH sensitive. The increase in pH rapidizes the gel formation process. This is due to the fact that whereas bases (OH^-) and acids (H^+) are both catalysts, they do so through various processes. It has been discovered that reaction rates are heavily influenced by pH, a number of pH-dependent rate profiles have been described in different literatures.

5.6 Solvent

The solvent has two crucial roles in the polymerization process as molecules are formed into nanoparticles. First, it must be able to hold the dissolved nanoparticles in the liquid so that they do not precipitate out of it. Secondly, it must play a role in aiding nanoparticle connectivity.

5.7 Temperature

Temperature impacts the gel duration because it accelerates the chemical kinetics of the many events involved in nanoparticle synthesis and the assembly of the nanoparticles in a gel network. Gelation is a gradual process that can take weeks or months at very low temperatures. The reactions that bind the nanoparticles to the gel network, on the other hand, happen so quickly at high temperatures that lumps form in their place, and a solid precipitates out of the liquid. To maximize the reaction time, the gelation temperature needs to be adjusted.

5.8 Time

The various phases in the gel formation process operate differently at various time scales depending on the sort of gel that is desired. In order to create a more homogeneous structure and a stronger gel, it is generally advised that the gel-forming process be delayed. Accelerating processes over a short period of time result in the formation of precipitates rather than a gel network, which can make a gel weak and hazy or prevent it from forming altogether.

5.9 Agitation

The mixing of the sol during gelation at this point should guarantee that the chemical reactions in the solution are produced equally, allowing all molecules to obtain an appropriate supply of the chemicals they need for these reactions to be carried out properly. In most cases, microscopic and macroscopic gel network domains are partially created throughout the liquid; however, agitation can occasionally disrupt the creation of these domains, causing the network fragments to reassemble into a larger network.

6. Advantages of the sol-gel method

One of the top methods for material synthesis so far is sol-gel technology. The use of the sol-gel process is quite beneficial for creating superior materials. Since the materials' introduction in 1960, the less expensive sol-gel processing method has been the most effective means to prepare target materials for both extensive and intensive study. By using the sol-gel process, metallic, inorganic, organic, and hybrid materials can be created. The sol-gel process is essential for producing a wide range of materials, from those needed for everyday usage to those used in extremely complex applications such as photonics, electronics, mechanics, biology, and medicine. The result of the sol-gel method is an improvement in the processing of traditional materials and their properties, as well as the synthesis of new materials. Due to its low-temperature nature, the organic-inorganic hybrids sol-gel technique is highly helpful for creating high-performance liquid chromatography. The following are some benefits of the sol-gel technique:

- Easy procedure.
- The creation of extremely pure products.

- The efficiency of synthesis is very high.
- Synthetic optical components have intricate shapes.
- Consistent synthesis of composite oxides.
- Customizing composition design and control for the synthesis of homogeneous materials.
- The use of fibers, aerogels, and materials with unique shapes.
- More thorough surface coverage.
- Synthesis of thin-layer amorphous materials.
- Synthesis of materials with specialized physical properties, including low thermal expansion coefficient, low UV absorption, and high optical transparency.
- Production of porous and dense materials using organic and polymeric compounds.
- Precursors have high chemical reactivity as a result of the solution phase process.
- The creation of low-cost, high-quality materials. Glass can be fabricated using the low-temperature sol-gel technique in a variety of shapes, from the most basic to the most complex.
- Synthesis at low temperature.
- Preparation of high-purity products.
- Very high production efficiency.
- Production of optical components with complex shapes.
- Synthesis of uniform compounds in the form of composite oxides.
- Can produce thin bond coating to provide excellent adhesion between the metallic substrate and the top coat.
- Can produce a thick coating to provide corrosion protection performance.
- Can easily shape materials into complex geometries in a gel state.
- Can produce high-purity products because the organometallic precursor of the desired ceramic oxides can be mixed, dissolved in a specified solvent, and hydrolyzed into a sol, and subsequently a gel, the composition can be highly controllable.
- Can have low-temperature sintering capability, usually 200–600°C.

- Can provide a simple, economic, and effective method to produce high-quality coatings.

7. Applications of sol-gel technology

- Thermal insulation.
- Acoustic insulation.
- Protective optical coatings.
- Lightweight materials.
- Tough ceramics.
- Membranes and microfilters.
- Nuclear waste storage.
- Ultra-fine powder abrasives.
- Encapsulation of biomolecules for controlled drug release.
- It can be used in ceramics manufacturing processes, as an investment casting material, or as a means of producing very thin films of metal oxides for various purposes.
- Sol-gel-derived materials have diverse applications in optics, electronics, energy, space, (bio) sensors, medicine (e.g., controlled drug release), and separation (e.g., chromatography) technology. One of the more important applications of sol-gel processing is to carry out zeolite synthesis.
- Other elements (metals, metal oxides) can be easily incorporated into the final product and the silicalite sol formed by this method is very stable.
- Other products fabricated with this process include various ceramic membranes for microfiltration, ultrafiltration, nanofiltration, preevaporation, and reverse osmosis.
- Different applications of sol-gel technology is shown in **Figure 9**.

8. Disadvantages of sol-gel processing

The loose dispersion of particle proportions and particle clustering are two drawbacks of this method. Despite all the advantages, sol-gel technology has certain restrictions.

- sluggish kinetics, poor wear resistance, and weak bond formation.

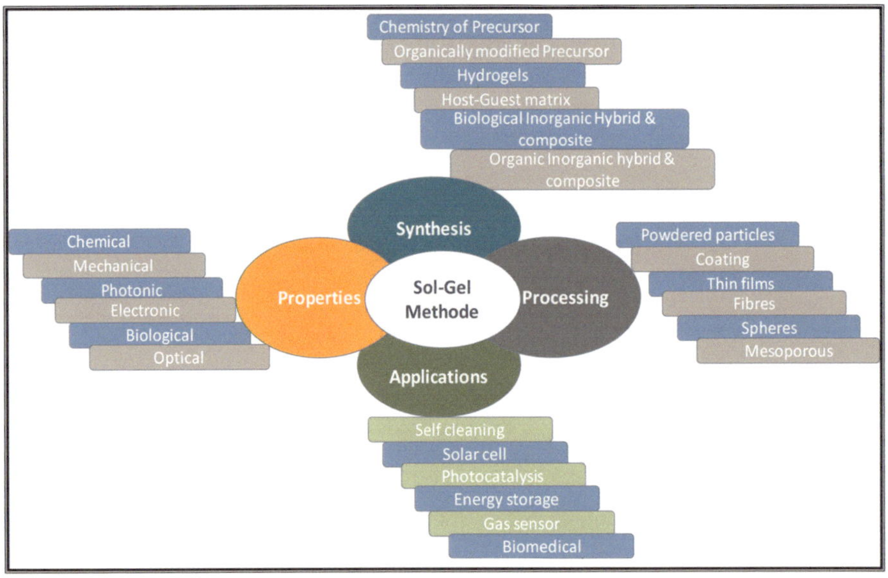

Figure 9.
Applications of sol-gel technology.

- The sol-gel method is a laborious and drawn-out process that requires extra care, particularly during times of drying and aging.

- Monoliths may shrink during the densification step, and this can lead to surface cracking brought on by capillary forces.

- Sol-gel materials cannot be employed in large-scale optical coatings since the required precursors are pricey and moisture-sensitive.

The property of the material will be diminished by the development of undesirable byproducts because the process involves numerous chemical reactions. Materials formed from sol-gel can be produced in a variety of shapes, including monoliths, powders, fibers, and films. Sol-gel materials can be categorized according to their uses as catalysts, inorganic pigments, pharmaceuticals, and magnetic and metallic nanoparticles, as well as materials used to encase biological molecules like proteins and enzymes. Films are crucial from a technological perspective, and in this review, we concentrate primarily on sol-gel-generated films that are produced *via* electro-spraying and dip-coating processes [38].

9. Summary

The science and technology of sol-gel have the ability to significantly alter the characteristics of materials. The fact that sol-gel science operates at room temperature and is insensitive to the atmosphere is one of its most notable advantages. These properties enable it to be used with a variety of substances that cannot withstand high temperatures and do not require the researcher to take extra. Any sol-gel system can have its water content, modifiers, and solvent changed, which changes the network's connectivity and impacts the characteristics. Sol-gel chemistry is used in a number of ongoing research. The sol-gel method is a popular and useful technique for producing nanoparticles with different chemical compositions. One of the widely utilized, adaptable materials in the field of nanotechnology is sol-gel-based nanoparticles, thin films, and fibers. A wide variety of applications have been extensively employed for the sol-gel-derived nanomaterials. This chapter largely focuses on the principles of sol-gel science with an emphasis on the fundamentals of creating nanomaterials and their many applications.

Author details

Satya Sopan Mahato[1], Disha Mahata[2], Sanjibani Panda[3] and Shrabani Mahata[3]*

1 Department of Electronics and Communication Engineering, National Institute of Science and Technology, Berhampur, India

2 Department of Electronics and Telecommunication Engineering, Kalinga Institute of Industrial Technology, Bhubaneswar, India

3 Department of Chemistry, National Institute of Science and Technology, Berhampur, India

*Address all correspondence to: shrabani.mahata@gmail.com

References

[1] Brinker CJ, Scherer G. Sol-Gel Science. In: The Physics and Chemistry of sol-Gel Processing. 1st ed. New York: Academic Press; 1989. p. 881

[2] Zhao D, Avnir D. Introductory chapter: A brief semblance of the sol-gel method in research. In: Sol-Gel Method. London, UK, London, UK: IntechOpen; 2018. DOI: 10.5772/intechopen.82487

[3] Mokhtari K, Salem S. A novel method for the clean synthesis of nanosized cobalt based blue pigments. RSC Advances. 2017;**17**(7):29899. DOI: 10.1016/j.watres.2008.08.0152

[4] Sakka S. The outline of applications of the sol–gel method. In: Klein L, Aparicio M, Jitianu A, editors. Handbook of sol-Gel Science and Technology. Cham: Springer; 2018. DOI: 10.1007/978-3-319-19454-7_53-1

[5] Bokov D, Jalil AT, Chupradit S, et al. Nanomaterial by sol-gel method: Synthesis and application. Advances in Materials Science and Engineering. 2021;**2021**:5102014. DOI: 10.1155/2021/5102014

[6] Karataş A, Algan AH. Template synthesis of tubular nanostructures for loading biologically active molecules. Current topics in medicinal chemistry. Current Topics in Medicinal Chemistry. 2017;**17**(13):1555-1563. DOI: 10.5772/intechopen

[7] Owens GJ, Singh RK, Foroutan F, Alqaysi M, Han C-M, Mahapatra C, et al. Sol–gel based materials for biomedical applications. Progress in Materials Science. 2016;**77**:1-79. DOI: 10.1016/j.pmatsci.2015.12.001

[8] Gill JK, Orsat V, Kermasha S. Screening trials for the encapsulation of laccase enzymatic extract in silica sol-gel. Journal of Sol-Gel Science and Technology. 2018;**85**:657-663. DOI: 10.1007/s10971-017-4575-9

[9] Hua L, Tan S N: Applications of a versatile sol-gel derived renewable electrode for capillary electrophoresis. Fresenius' Journal of Analytical Chemistry. Au 2000;**367**(8):697-700. DOI: 10.1007/s002160000477

[10] Mondal B, Usha K, Mahata S, Kumbhakar P, Nandi MM. Synthesis and characterization of nanocrystalline TiO_2 thin film for use as photoelectrodes in dye sensitized solar cell application. Transactions of the Indian Ceramic Society. 2011;**70**(3):173-177. DOI: 10.1080/0371750X.2011.10600167

[11] Nahak BK, Subudhi TSK, Pradhan LK, Panigrahi A, Roshan R, Mahato SS, et al. An investigation on photocatalytic dye degradation of Rhodamine-6G dye with Fe- and Ag-doped TiO_2 thin films. Part of the Lecture Notes in Electrical Engineering book series (LNEE). 2020; **673**:295-307. DOI: 10.1007/978-981-15-5546-6_24

[12] Giordano C, Antonietti M. Synthesis of crystalline metal nitride and metal carbide nanostructures by sol–gel chemistry. NanoToday. 2011;**6**(4): 366-380. DOI: 10.1016/j.nantod.2011.06.002-2011

[13] Seifitokaldani A, Savadogo O, Perrier M. Stability and catalytic activity of titanium oxy-nitride catalyst prepared by in-situ urea-based sol–gel method for the oxygen reduction reaction (ORR) in acid medium. International Journal of Hydrogen Energy. 2015;**40**(33): 10427-10438. DOI: 10.1016/j.ijhydene.2015.06.002-2015

[14] Soria AC, Sánchez JL, Miralles CG, Varela M, Navarro E, González C, et al. Novel one-pot sol-gel synthesis route of Fe_3C/few-layered graphene core/shell nanoparticles embedded in a carbon matrix. Journal of Alloys and Compounds. 2022;**902**(5):163662. DOI: 10.1016/j.jallcom.2022.163662-2022, 902 (5) 163662

[15] Lee BS, Lin HP, Chan JCC, Wang WC, Tsai YH, Lee YL. A novel sol-gel derived calcium silicate cement with short setting time for application in endodontic repair of perforations. International Journal of Nanomedicine. 2018;**13**:261-271. DOI: 10.2147/IJN. S150198

[16] Zhang P, Bi L, Zhang D, Wang X, Wang W, Lei X, et al. Synthesis of Fe_3C branches via a hexamethylenetetramine route. Materials Research Bulletin. 2016; **76**:327-331. DOI: 10.1016/j. materresbull.2015.12.038-2016

[17] Shi C, Ding GS, Tang AN, Qiao YY. Synthesis and evaluation of ion-imprinted sol-gel material of selenite. Analytical Methods. 2017;**9**:1658-1664

[18] Mahata S, Mahato SS, Usha K, Mandal N, Mukherjee K, Mondal B. Chemical modification of titanium isopropoxide for producing stable dispersion of titanianano-particles. Materials Chemistry and Physics. 2015; **151**:267-274. DOI: 10.1016/j.match emphys.2014.11.065

[19] Mahata S, Nandi MM, Mondal BN. Preparation of high solid loading titania suspension in gelcasting process using boil rice extract as binder. Ceramics International. 2012;**38**(2):909-918

[20] Jhariya N, Roshan R, Mahato SS, Mahata S. Hydrothermal synthesis of surface functionalized semiconducting Nano crystals and study of their photo induced interaction with natural dye. International Research Journal of Engineering and Technology. 2018;**5**(5): 3892-3895

[21] Mahata S, Kundu D. Hydrothermal synthesis of aqueous nano-TiO2 sols. Materials Science-Poland. 2009;**27**(2): 463-470

[22] Vinogradov AV, Vinogradov VV. Low-temperature sol–gel synthesis of crystalline materials. RSC Advances. 2014;**4**:5903-45919. DOI: 10.1039/ C4RA04454A

[23] Borlaf M, Moreno R. Colloidal sol-gel: A powerful low-temperature aqueous synthesis route of nanosized powders and suspensions. Open Ceramics. 2021;**8**:100200. DOI: 10.1016/ j.oceram.2021.100200

[24] Chotirat L, Niyomwas S, Wongpisan W, Supothina S. Low-temperature synthesis of vanadium dioxide thin films by sol-gel dip coating method. Journal of Nanotechnology. 2021;7:4868152. DOI: 10.1155/2021/ 4868152

[25] Hudson MJ, Peckett JW, Harris PJF. Low-temperature sol–gel preparation of ordered nanoparticles of tungsten carbide/oxide. Industrial & Engineering Chemistry Research. 2005;**44**(15): 5575-5578. DOI: 10.1021/ie040247v

[26] Borilo L, Kozik V, Vorozhtsov A, Klimenko V, Khalipova O, Agafonov A, et al. The low-temperature sol-gel synthesis of metal-oxide films on polymer substrates and the determination of their optical and dielectric properties. Nanomaterials. 2022;**12**(23):4333. DOI: 10.3390/ nano12234333

[27] Kozhevnikova NS, Ul'yanova ES, Shalaeva EV, et al. Low-temperature sol–

gel synthesis and Photoactivity of Nanocrystalline TiO_2 with the Anatase/ Brookite structure and an amorphous component. Kinetics and Catalysis. 2019; **60**:325-336. DOI: 10.1134/ S002315841903008X

[28] Reiser JT, Ryan TV, Wall NA. Sol–gel synthesis and characterization of gels with compositions relevant to hydrated glass alteration layers. ACS Omega. 2019; **4**(15):16257-16269. DOI: 10.1021/ acsomega.9b00491

[29] Yarbrough R, Davis K, Dawood S, Rathnayake H. A sol–gel synthesis to prepare size and shape-controlled mesoporous nanostructures of binary (II–VI) metal oxides. RSC Advances. 2020;**10**:14134-14146. DOI: 10.1039/ D0RA01778G

[30] Tripathi VK, Nagarajan R. Sol–gel synthesis of high-purity actinide oxide ThO_2 and its solid solutions with technologically important tin and zinc ions. Inorganic Chemistry. 2016;**55**(24): 12798-12806. DOI: 10.1021/acs. inorgchem.6b02086

[31] Mateos D, Valdez B, Castillo JR, Nedev N, Curiel M, Perez O, et al. Synthesis of high purity nickel oxide by a modified sol-gel method. Ceramics International. 2019;**45**(9):11403-11407. DOI: 10.1016/j.ceramint.2019.03.005

[32] Lazareva SV, Shikina NV, Tatarova LE, Ismagilov ZR. Synthesis of high-purity silica nanoparticles by sol-gel method. Eurasian Chemico-Technological Journal. 2017;**19**:295-302. DOI: 10.18321/ectj677

[33] Yang D, Yang G, Liang G, Guo Q, Li Y, Li J. High-surface-area disperse silica nanoparticles prepared via sol-gel method using L-lysine catalyst and methanol/water co-solvent. Colloids and Surfaces A: Physicochemical and Engineering Aspects. 2021;**610**:125700. DOI: 10.1016/j.colsurfa.2020.125700

[34] Bangi UKH, Park CS, Baek S, Park HH. Sol–gel synthesis of high surface area nanostructured zirconia powder by surface chemical modification. Powder Technology. 2013; **239**:314-318. DOI: 10.1016/j. powtec.2013.02.014

[35] Periyat P, Laffir F, Tofaila SAM, Magner E. A facile aqueous sol–gel method for high surface area nanocrystalline CeO_2. RSC Advances. 2011;**1**:1794-1798. DOI: 10.1039/ C1RA00524C

[36] Harmer MA, Vega AJ, Flippen RB. Sol-gel synthesis of high surface area aluminum phosphate: A thermally reversible sol-gel system. Chemistry of Materials. 1994;**6**(11):1903-1905. DOI: 10.1021/cm00047a001

[37] Aronne A, Turco M, Bagnasco G, Pernice P, Serio MD, Clayden NJ, et al. Synthesis of high surface area Phosphosilicate glasses by a modified sol–gel method. Chemistry of Materials. 2005;**17**(8):2081-2090. DOI: 10.1021/ cm047768t

[38] Khalil T, El-Nour FA, El-Gammal B, Boccaccini AR. Determination of surface area and porosity of sol-gel derived ceramic powders in the system TiO_2–SiO_2–Al_2O_3. Powder Technology. 2001; **114**(1):106-111. DOI: 10.1016/ S0032-5910(00)00271-0

[39] Hassan AF, Abdelghny AM, Elhadidy H, et al. Synthesis and characterization of high surface area nanosilica from rice husk ash by surfactant-free sol–gel method. Journal of Sol-Gel Science and Technology. 2014;**69**:465-472. DOI: 10.1007/ s10971-013-3245-9

[40] Mukundan LM, Nirmal R, Vaikkath D, Nair PD. A new synthesis route to high surface area sol gel bioactive glass through alcohol washing: A preliminary study. Biomatter. 2013; **3**(2):e24288. DOI: 10.4161/biom.24288

[41] Islam MA, Sato T, Ara F, Basith MA. Sol-gel based synthesis to explore structure, magnetic and optical properties of double perovskite Y_2FeCrO_6 nanoparticles. Journal of Alloys and Compounds. 2023;**944**: 169066. DOI: 10.1016/j.jallcom.2023.169066

[42] Belkhiria F, Rhouma FIH, Gammoudi H, Amlouk M, Chtourou R. Structural and optical investigations on LaGaO3 synthesis by sol–gel method along with a blue luminescence property. Inorganic Chemistry Communications. 2022;**143**:109807. DOI: 10.1016/j.inoche.2022.109807

[43] Raimundo RA, Silva VD, Ferreira LS, Loureiro FJA, Fagg DP, Macedo DA, et al. High magnetic moment of nanoparticle-sphere-like Co, Fe based composites and alloys prepared by proteic sol-gel synthesis: Structure, magnetic study and OER activity. Journal of Alloys and Compounds. 2023; **940**:168783. DOI: 10.1016/j.jallcom.2023.168783

[44] Skruodiene M, Juodvalkyte R, Inkrataite G, Pakalniskis RAR, Sarakovskis A, Skaudzius R. Sol-gel assisted molten-salt synthesis of novel single phase Y_3–2xCa_2xTaxAl$_5$–xO12:1% Eu garnet structure phosphors. Journal of Alloys and Compounds. 2022;**890**: 161889. DOI: 10.1016/j.jallcom.2021.161889

[45] Carrera-Figueiras C, Pérez-Padilla Y, Alejandro Estrella-Gutiérrez M, Uc-Cayetano G, Juárez-Moreno JA, Avila-Ortega A. Surface science engineering through sol-gel process. In: Applied Surface Science. London, UK: IntechOpen; 2019. DOI: 10.5772/intechopen.83676

[46] Pilliar RM. Surface Coating and Modification of Metallic Biomaterials. Cambridge, UK: Woodhead Publishing; 2015. pp. 185-217. DOI: 10.1016/B978-1-78242-303-4.00006-5

[47] Textor T. 9 - modification of textile surfaces using the sol-gel technique. In: Surface Modification of Textiles. Cambridge, UK: Woodhead Publishing; 2009. pp. 185-213. DOI: 10.1533/9781845696689.185

[48] Kane N, Zhou Y, Zhang W, Ding Y, Luo Z, Hua X, et al. Precision surface modification of solid oxide fuel cells via layer-by-layer surface sol–gel deposition. Journal of Materials Chemistry A. 2022; **10**:8798-8806. DOI: 10.1039/D2TA00458E

[49] Tranquillo E, Bollino F. Surface modifications for implants lifetime extension: An overview of sol-gel coatings. Coatings. 2020;**10**(6):589. DOI: 10.3390/coatings10060589

[50] Vihodceva S, Kukle S. Cotton fabric surface modification by sol-gel deposition of ZnO thin films. IOP Conference Series: Materials Science and Engineering. 2012;**38**:012022. DOI: 10.1088/1757-899X/38/1/012022

[51] Kashouty MA, Elsayad H, Salem T, Elhadad S. Egyptian Journal of Chemistry. 2020;**63**(9):3301-3311. DOI: 10.21608/ejchem.2020.24441.2464

[52] Bollino F, Tranquillo E. Surface modification of implants by sol-gel coating technology: Advantages and applications. Materials Proceedings. 2020;**2**:7. DOI: 10.3390/CIWC2020-06827

[53] Bednorz JG, Mueller KA. Possible high Tc superconductivity with barium-lanthanum-copper-oxygen system. Zeitschrift für Physik. 1986;**B-64**: 189-193

[54] Bernardino S, Estrela N, Ochoa-Mendes V, Fernandes P, Fonsseca LP. Optimization in the immobilization of penicillin G acylase by entrapment in xerogel particles with magnetic properties. Journal of Sol-Gel Science and Technology. 2011;**58**:545-556

[55] Boettcher H, Slowak P, Suss W. Sol–gel carrier systems for controlled drug delivery. Journal of Sol-Gel Science and Technology. 1998;**1**(3):277-281

[56] Boninsegna S, Dal TR, Dal MR, Carturan G. Alginate microsphere loaded with animal cells and coated by a siliceous layer. J sol–gel. Science and Technology. 2012;**61**:570-576

[57] Bono MS, Anderson AM, Caroll MK. Alumina aerogels prepared via rapid supercritical extraction. Journal of Sol-Gel Science and Technology. 2010;**53**: 16-22

[58] Boyle TJ, Buchheit CD, Rodriguez MA, Al-Shareef HN, Hernandez BA, Scott B, et al. Formation of SrBi2Ta2O9: Part I. synthesis and characterization of a novel sol–gel solution for production of ferroelectric SrBi2Ta2O9 thin films. Journal of Materials Research. 1996;**11**:2274-2281

[59] Hu A, Yao Z, Yu X. Phase behavior of a sodium dodecanol allyl sulfosuccinic di-ester/n-pentanol/methyl acrylate/butyl acrylate/water microemulsion system and preparation of acrylate latexes by microemulsion polymerization. Journal of Applied Polymer Science. 2009;**113**:2202-2208. DOI: 10.1002/app.30284

[60] Pileni M-P. Marie-Paule, Pileni, nanocrystals: Fabrication, organization and collective properties. Comptes Rendus Chimie. 2003;**6**(8–10):965-978. DOI: 10.1016/j.crci.2003.07.007

[61] Pileni MP. Mesostructured fluids in oil-rich regions:Structural and Templating approaches. Langmuir. 2001;**17**(24):7476-7486. DOI: 10.1021/la010538y

[62] Pileni MP. Control of the size and shape of inorganic nanocrystals at various scales from Nano to macrodomains. The Journal of Physical Chemistry C. 2007;**111**(26):9019-9038. DOI: 10.1021/jp070646e

[63] Pileni MP. Supracrystals of inorganic nanocrystals: An open challenge for new physical properties. Accounts of Chemical Research. 2008;**41**:1799-1809. DOI: 10.1021/ar800082q

[64] Hoffmann MR, Martin ST, Choi W, Bahnemann DW. Environmental applications of semiconductor photocatalysis. Chemical Reviews. 1995; **95**:69-96. DOI: 10.1021/cr00033a004

[65] Linsebigler AL, Lu G, Yates JT. Photocatalysis on TiO_2 surfaces: Principles, mechanisms, and selected results. Chemical Reviews. 1995;**95**: 735-758. DOI: 10.1021/cr00035a013

[66] Dvoranova D, Brezova V, Mazur M, Malati MA. Investigations of metal-doped titanium dioxide photocatalysts. Applied Catalysis. B, Environmental. 2002;**37**:91-105. DOI: 10.1016/S0926-3373(01)00335-6

[67] Iwasaki M, Hara M, Kawada H, Tada H, Ito S. Cobalt iondoped TiO_2 photocatalyst response to visible light. Journal of Colloid and Interface Science. 2000;**224**:202-204. DOI: 10.1006/jcis.1999.6694

[68] Sakthivel S, Shankar VM, Palanichamy M, Arabindoo B, Bahnemann DW, Murugesan V. Enhancement of photocatalytic activity by metal deposition: Characterisation and photonic efficiency of Pt, Au and Pd deposited on TiO_2 catalyst. Water Research. 2004;**38**:3001-3008. DOI: 10.1016/j.watres.2004.04.046

[69] Gomez R, Bertin V, Lopez T, Schifter I, Ferrat G. Pt-Sn/Al_2O_3 sol-gel catalysts: Catalytic properties. Journal of Molecular Catalysis A. 1996;**109**:55-66. DOI: 10.1016/1381-1169(95)00290-1

[70] Chilibon I, Marat-Mendes JN. Ferroelectric ceramics by sol-gel methods and applications: A review. Journal of Sol-Gel Science and Technology. 2012;**64**:571-671. DOI: 10.1007/s10971-012-2891-7

[71] Fukushima J, Kodaira K, Matsushita T. Preparation of ferroelectric PZT films by thermal decomposition of organometallic compounds. Journal of Materials Science. 1984;**19**:595-598. DOI: 10.1007/BF02403247

[72] Quek HM, Yan MF. Sol-gel preparation and dielectric properties of lead iron niobate thin films. Ferroelectrics. 1987;**74**:95-108

[73] Vest RW, Xu J. Preparation and properties of PLZT films from metallo-organic precursors. Ferroelectrics. 1989;**93**:21-29. DOI: 10.1080/00150198908017316

[74] Melnick BM, Cuchiaro JD, McMillian LD. Paz de Araujo C a, Scott J F: Process optimization and characterization of device worthy sol-gel based PZT for ferroelectric memories. Ferroelectrics. 1990;**112**:329-351. DOI: 10.1080/00150199008008237

[75] Spierings GACM, Uleanaers MJE, Kampschoeer GLM, van Hal HAM, Larsen PK. Preparation and ferroelectric properties of $PbZr_{0.53}Ti_{0.47}O_3$ thin films by spin coating and metalorganic decomposition. Journal of Applied Physics. 1991;**70**:2290-2296. DOI: 10.1063/1.349422

[76] Vest RW, Zhu W. Films of 60/40 PZT by the MOD process for memory applications. Ferroelectrics. 1991;**119**:61-75. DOI: 10.1080/00150199108223327

[77] Xu Y, Chen J, Xu R, Mackenzie JD. The self-biased heterojunction effect of ferroelectric thin film on silicon substrate. Journal of Applied Physics. 1990;**67**:2985-2991. DOI: 10.1063/1.345420

[78] Kung HH, Ko EI. Preparation of oxide catalysts and catalyst supports—A review of recent advances. Chemical Engineering Journal. 1996;**64**:203-214. DOI: 10.1016/S0923-0467(96)03139-9

[79] Stuky GD, Marder SR, Sohn JE. Materials for Nonlinear Optics. Washington D.C.: Am. Chem. Soc.; 1991. DOI: 10.1021/bk-1991-0455

[80] Mackenzie JD, Xu R, Yuhuan X. In: Hench LL, West JK, editors. Chemical Processing of Advanced Materials. New York: John Wiley & Sons; 1992. p. 365. DOI: 10.1021/bk-1991-0455

[81] Mackenzie JD. Nonlinear optical materials by the sol-gel method. Journal of Sol-Gel Science and Technology. 1993;**1**:7-19. DOI: 10.1007/BF004864255

[82] Hench LL, Wilson MJR, Non-Cryst J. Processing of gel-silica monoliths for optics: Drying behavior of small pore gels. Journal of Non-Crystalline Solids. 1990;**121**(1–3):234-243. DOI: 10.1016/0022-3093 (90)90138-C

[83] Komarneni S, Ravindranathan P, Bhalla AS, et al. Solution-sol-gel processing of superconductors. Bulletin of Materials Science. 1991;**14**:359-365. DOI: 10.1007/BF02747342

[84] Hussein AAA, Hasan N, Adi M, Hussein A, Mohammed FQ. Synthesis of bulk superconductors using the sol-gel method. Conference Proceedings. 2022; **2437**:020051. DOI: 10.1063/5.0108818

[85] Hazen RM, Prewitt CT, Angel RJ, Ross NL, Finger LW, Hadidiacos CG, et al. Superconductivity in the high-T_c Bi-Ca-Sr-Cu-O system: Phase identification. Physical Review Letters. 1988;**60**:1174. DOI: 10.1103/PhysRevLett.60.1174

[86] Kordas G. Sol-gel processing of ceramic superconductors. Journal of Non-Crystalline Solids. 1990;**121**(1–3): 436-442. DOI: 10.1016/0022-3093(90)90172-I

[87] Zhao X, Wang T. Songchol Hong, Sun D, Wang N, Chae G and qi: Superconducting properties and non-isothermal crystallization kinetics of a $Bi_2Sr_2CaCu_2O_{8+\delta}$ (Bi2212) superconductor prepared by the Pechini sol–gel method. RSC Advances. 2019;**9**: 35280-35288. DOI: 10.1039/C9RA07586K

[88] Zalga A, Beganskienė A, Kareiva A. Sol-gel synthesis and superconducting properties of Bi-2212 high-T-C superconductors. Polish Journal of Chemistry. 2007;**81**:1547-1553

[89] Kurniawan B et al. Synthesis of superconductor La1.84Sr0.16CuO4 with sol-gel method. Journal of Physics: Conference Series. 2019;**1170**:012052. DOI: 10.1088/1742-6596/1170/1/012052

[90] Mathur S, Shen H, Lecerf N, Jilavi MH. Gel synthesis route for the preparation of $Y(Ba_1{-}xSr_x)2Cu_4O_8$ superconducting oxides. Journal of Sol-Gel Science and Technology. 2002;**24**:57-68. DOI: 10.1023/A: 1015113616733

[91] Zhang JM, Marcy HO, Tonge LM, Wessels BW, Marks TJ, Kannewurf CR. Organometallic chemical vapor deposition of superconducting, high Tc Pb-doped Bi-Sr-Ca-Cu-O thin films. Applied Physics Letters. 1989;**55**:1906. DOI: 10.1063/1.102330

[92] Braun S, Shtelzer S, Rappoport S, Avnir D, Ottolenghi M. Biocatalysis by sol–gel entrapped enzymes. Journal of Non-Crystalline Solids. 1992;**147–148**: 739-748. DOI: 10.1016/S0022-3093(05)80708-2

[93] Brown P, Hope-Weeks LJ. The synthesis and characterization of zinc ferrite aerogels prepared by epoxide addition. J sol–gel. Science and Technology. 2009;**51**:238-243. DOI: 10.1007/s10971-009-1985-3

[94] Brown LM, Mazdiyasni KS. Cold-pressing and low-temperature sintering of alkoxy-derived PLZT. Journal of the American Ceramic Society. 1972;**55**: 541-542. DOI: 10.1111/j.1151-2916.1972.tb13432.x

[95] Campostrini R, Carturan G, Caniato R, Piovan A, Filippini R, Innocenzi G, et al. Immobilization of plant cell in hybrid sol–gel materials. J sol–gel. Science and Technology. 1990;7: 87-97. DOI: 10.1007/BF00401888

[96] Gill I, Ballesteros A. Encapsulation of biologicals within silicate, siloxane, and hybrid sol–gel polymers: An efficient and generic approach. Journal of the American Chemical Society. 1998; **120**(34):8587-8598. DOI: 10.1021/ja9814568

[97] Available from: https: //www.isgs.org/community/2020-in-review-the-achievements-of-the-sol-gel-science-and-technology-community/

[98] Parashar M, Shukla VK, Singh R. Metal oxides nanoparticles via sol–gel method: A review on synthesis, characterization and applications. Journal of Materials Science: Materials in Electronics. 2020;**31**:3729-3749. DOI: 10.1007/s10854-020-02994-8

[99] Thiagarajan S, Sanmugam A, Vikraman D. Facile methodology of sol-gel synthesis for metal oxide nanostructures. Recent applications in sol-gel synthesis. In: Chandra U, editor. Recent Applications in sol-Gel Synthesis. London, UK: IntechOpen; 2017. DOI: 10.5772/intechopen.68708

[100] Niederberger M. Nonaqueous Sol–Gel Routes to Metal Oxide Nanoparticles. Accounts of Chemical Research. 2007;**40**(9):793-800. DOI: 10.1021/ar600035e

[101] Mitra A, De G. Chapter 6-sol-gel synthesis of metal nanoparticle incorporated oxide films on glass. In: Karmakar B, Rademann K, Stepanov AL, editors. Glass Nanocomposites. New York, USA: William Andrew Publishing; 2016. pp. 145-163. DOI: 10.1016/B978-0-323-39309-6.00006-7

Section 2

Synthesis of Materials Using Sol-Gel

Chapter 3

TiO_2 Nanostructures by Sol-Gel Processing

Srinivasa Raghavan

Abstract

This book chapter discusses the versatile sol-gel processing technique that has been used to synthesize the nanostructures of titanium dioxide (TiO_2) and their different morphologies. The sol-gel syntheses of different nanostructures of TiO_2, namely TiO_2 nanoparticles, nanocrystalline thin film, nanorods, nanofibers, nanowires, nanotubes, aerogels, and opals are described. These nanostructures have been characterized by Scanning Electron Microscopy (SEM) and Transmission Electron Microscopy (TEM) whose images clearly depict the formation of the nanostructures. Some of the morphologies of nano-TiO_2 such as nanorods, nanotubes, nanofibers, nanowires, have been synthesized by sol-gel process in combination with spin-coating, dip-coating, template, surfactant, diblock polymer, micelles, polystyrene. In comparison to the bulk TiO_2, presence of porous and nanocrystalline morphologies has played a role in enhancing the performance in applications such as photovoltaics, photocatalysis, photocatalytic water-splitting, H_2 storage, gas sensors, photochromic, opto-electronic, and electrochromic devices. The chapter concludes with challenges and practical concerns in using the sol-gel process to produce thin films of complex oxides, porous nanostructures, solid nanorods, nanotubes, which need to be addressed in future research efforts.

Keywords: sol-gel process, TiO_2 nanostructures, sol-gel films, electrophoretic deposition, template-filling, nanorods, aerogels, opals, nanotubes, nanowires

1. Introduction

Titanium dioxide (aka titania, TiO_2) was discovered by English Mineralogist William Gregor (1791) in black magnetic sand in Cornwall. Few years later, it was independently isolated from the mineral Rutile and named by German scientist M.H. Klaproth in 1795. It was commercially produced during 1920s, as pigment [1]. Subsequently, it was used in sunscreens [2], paints [3], ointments, toothpaste [4], and so on.

Titanium dioxide is a colorless, opaque, chemically inert, non-toxic, and a semiconducting material that shows photocatalytic activity upon exposure to the light of energy beyond its bandgap value (3–3.2 eV). It shows absorption only in UV region. TiO_2 occurs in nature as three crystalline forms: Rutile, Anatase, and Brookite. Rutile phase shows absorption at slightly higher wavelength than the anatase form and the latter has been studied for its optical properties.

Ever since Fujishima—Honda effect of splitting of water using a TiO_2 electrode under UV light was discovered in 1972 [5–7], research efforts on TiO_2 have gained momentum in the past decades, especially in areas of photocatalysis, photovoltaics, photoelectrochemical cells, environmental pollution control, and sensors [8–11].

There has been enormous research activity on nanoscience and nanotechnology in the past decades. When a bulk material is brought down to smaller size, and further to nanoscale, there has been a paradigm shift in the material's physical and chemical properties. The nanomaterials surface area increases with decreasing size of the nanomaterial. Ability of these nanomaterials to transport electrons/holes faster in presence of light makes it attractive for photocatalytic/PV applications [12–15]. With the advent of nanoscience and nanotechnology, breakthroughs were made in the approaches to syntheses/modifications of TiO_2 nanomaterials. Thus, we could obtain newer nanomaterials with different morphologies such as nanoparticles, nanocrystalline films, nanorods, nanotubes, aerogels, opals, nanowires, as well as mesoporous, and photonic structures. These new nanomaterials exhibit optical, structural, electronic, and thermal properties that are largely determined by their size and shape [16].

Enormous surface area of the nanocrystalline TiO_2 compared to the bulk titania, is critical to applications in adsorption, catalysis, sensing, H_2 storage, photovoltaics, and wastewater remediation. Nanocrystalline titania has been widely investigated for its photocatalytic activity that was used to mitigate water pollution by means of either (i) removal of pollutants by adsorption, or (ii) photodegradation of organic dyes/drugs in industrial and domestic wastewater.

The bandgap of TiO_2 bulk phase lies in the UV region (3.0 eV for rutile phase and 3.2 eV for anatase phase). You may recall that the bandgap of a semiconductor is the energy difference between higher-energy levels called Conduction Band (CB) and the lower-lying Valence Band (VB). As the particle size decreases, energy levels tend to become more discrete, thereby increasing the bandgap of the nanomaterial. Such a widening of the bandgap in nanomaterials has been attributed to the quantum confinement effect [17, 18]. Sakai et al. [19] found that the lower edge of the conduction band for the TiO_2 nanosheet was approximately 0.1 V higher, while the upper edge of the valence band was 0.5 V lower than that of the bulk (anatase) TiO_2. Thus, the nanocrystalline titania is transparent to the visible light since it has a wide bandgap (4–4.5 eV). In a PV device, absorption of UV (high energy) light by TiO_2 promotes electrons to the excited state which must be transported quickly to the cathode through a load (an electric lamp) to complete the circuit. This promotion of electrons in TiO_2 can be done using visible light (i.e., using less energy) by enhancing its visible absorption edge. It is done by adsorbing a colored dye molecule that shows maximum absorption in visible region. The adsorbed dye molecule produces photogenerated electrons on absorption of light of less energy (visible light) compared to a pure TiO_2 electrode. Thus-generated electrons can either be used to generate electric current in a PV device or can be used in a photochromic glasses, or in a photoelectrochromic displays, or can be involved in the photodegradation of pollutants in water. This chapter presents only the synthetic details pertaining to the different TiO_2 nanostructures, particularly by sol-gel processing and some combination techniques where applicable. Discussion on those TiO_2 nanostructures that are formed by other synthetic techniques is beyond the scope of the chapter.

2. Synthetic methods for TiO_2 nanostructures

Several wet chemical methods are available for the syntheses of nanomaterials that include sol-gel processing, The choice of the synthetic technique largely depends on the nanostructures and morphologies of the materials that are desired. However, sol-gel process is so versatile a technique that it offers a possibility of further processing of the sol, using an appropriate material/template to obtain the desired nanostructures (**Figure 1**) which include thin films, nanofiber, nanowire, nanotube, nanorod, aerogel, opal, microspheres, nanocrystal, doped nanoparticles, etc.

2.1 Sol-gel method

Sol-gel processing has evolved as a powerful, yet versatile technique for the syntheses of inorganic materials such as glasses and ceramics [20–22]. It is a wet chemical process in which soluble metal alkoxide or nitrate precursor material is hydrolyzed to form a colloidal dispersion called sol. Subsequently, the sol undergoes polycondensation and forms infinite network of particles called gel. This is followed by aging which completes polymerization and removal of solvent converts the liquid sol into solid gel.

Further processing of the sol-gel by spin-coating or dip-coating on a substrate yields thin films. The sol is a low-viscosity liquid and with time, colloidal particles bond to form 3-dimensional network called gel. During gelation, viscosity increases sharply. With appropriate control of the viscosity of the sol, it can be spun into nanofibers, or can be grown into nanorods or nanotubes by passing the sol through the pores of a pre-formed template (membranes of alumina or polycarbonate (PC) with pores).

The sol-gel process has advantages over traditional methods: (i) high-purity materials can be synthesized at lower temperature; (ii) homogeneous multicomponent systems such as doped materials can be prepared by mixing appropriate

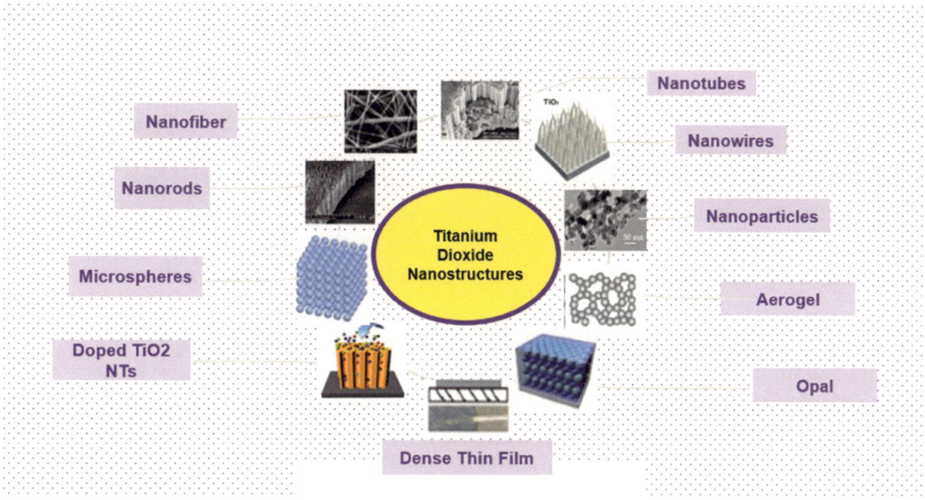

Figure 1.
Representative nanostructures of titanium dioxide.

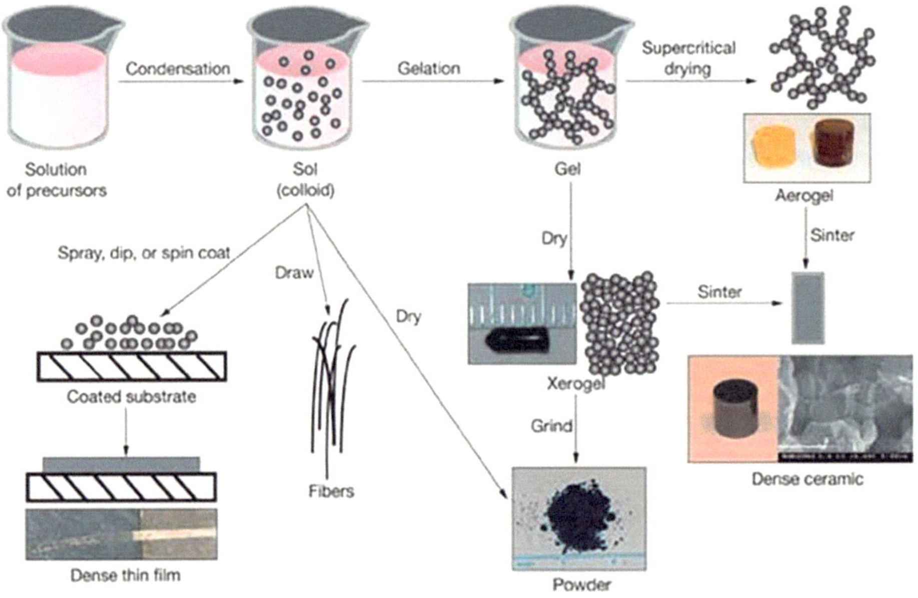

Figure 2.
Steps involved in sol-gel processing. Adapted from the book entitled "Introductory Chapter: A brief Semblance of Sol-gel Method in Research" by GV Aguilar. IntechOpen. 2018. DOI: 10.5772/intechopen.82487.

precursor solutions; (iii) sol-gel process has been used to prepare nanostructures with different morphologies of the desired semiconductor and other inorganic materials; (iv) metal oxides and ceramics are chemically and thermally stable and this precludes the use of conventional methods (PVD, CVD) to synthesize nano-structures of the metal oxides.

Sol-gel steps (**Figure 2**) are described below:

Sol-gel monoliths are made, in general, by three approaches: (method 1) gelation of solution of colloidal powders; (method 2) hydrolysis and polycondensation of metal alkoxide or nitrate precursors, followed by hypercritical drying of gels; and (method 3) hydrolysis and polycondensation of alkoxide precursors followed by aging and drying under ambient conditions.

After gelation, the pore liquid is removed from the gel network by hypercritical drying, the network does not collapse, and a low-density *Aerogel* is formed.

Processing steps of sol-gel are: (i) mixing; (ii) casting; (iii) gelation; (iv) aging; (v) drying; (vi) dehydration or chemical stabilization; (vii) densification.

i. *Mixing*: the liquid Ti-alkoxide precursor is hydrolyzed by mixing in water at a pH in which it is not precipitated. This reaction yields $Ti(OH)_4$ which undergoes condensation forming ≡Ti−O−Ti≡ bonds. Further polycondensation brings about additional linkage of ≡Ti−OH tetrahedra and this eventually, forms −Ti−O−Ti− network. Water and alcohol formed in the condensation reactions remain in the pores of the network. Hydrolysis and polycondensation reactions continue to occur, and this results in the formation of interconnected −Ti−O−Ti− bonds in solution, that is, Sol is formed. The size of the sol particles depend on pH and the ratio $[H_2O]/[Ti(OR)_4]$.

ii. *Casting*: since the sol is a low viscous liquid, it can be cast into a mold, or it can be drawn into wires, rods, tubes, fibers, ...

iii. *Gelation*: with time, the colloidal particles get interconnected to form a 3-D network, called Gel. With gelation, viscosity of the sol increases sharply and with a proper control of viscosity, fibers can be spun.

iv. *Aging*: aging of the gel (called *Syneresis*) refers to maintaining the cast object, for hours to days, completely immersed in liquid. During aging, polycondensation continues, the thickness of interparticle necks increases, which lowers the porosity and increases the strength of the gel. The strength of the gel increases with aging time.

v. *Drying*: the liquid in the pores of the gel is removed during drying. For the smaller pores (<20 nm), gel starts cracking due to high surface energy. This cracking can be prevented by minimizing the liquid surface energy either by hypercritical drying (method 2), or by using surfactants or by eliminating smaller pores (method 1), or by obtaining monodisperse pores by controlling rates of hydrolysis and condensation (method 3).

vi. *Dehydration*: ultraporous, stable solid is obtained by the removal of surface (Ti−O) bonds. Thus, the obtained stable, strong, and porous gel is optically transparent which finds applications as optical components, when impregnated with optically active Fluors, dyes or nonlinear polymers [23, 24].

vii. *Densification*: Heating the porous gel at high temperatures, removes the pores and causes densification to occur. For example, alkoxide-derived silica gels have been densified at temperature as low as 1000°C [25], and these gels possess superior purity and homogeneity, compared to other commercial processes.

Discussion on the sol-gel syntheses of different nanostructures of Titania such as Nanocrystalline thin films, nanoparticles, aerogels, opals, nanotubes, nanofibers, nanowires, microspheres, and nanorods, is presented in following sections.

2.1.1 Synthesis of titania thin films

Titania nanostructures have widely found applications in various fields which include gas sensing, photocatalysis, water purification, protein separation, Dye-sensitized solar cells (DSSC), solid-state DSSC, and perovskite solar cells (PSC) [26–31]. These solar cell devices could not achieve photon-to-current efficiency or the device stability, when compared with the conventional Si-based solar cells. However, the fabrication of Si-based solar cells involves expensive and energy-intensive processes. Therefore, large-scale fabrication of the present-day PV devices has been largely determined by the key factors: cost- and energy-efficiency. In this regard, research on the devices that were utilizing titania nanostructures have gained significant attention [32].

Earlier, Spray deposition was found to be a promising method to produce thin films for the PV devices [33]. Among the wet chemical synthetic methods of titania nanostructures, polymer (polystyrene-b-poly (ethene oxide)) template-assisted sol-gel process in combination with spray deposition has been found to be powerful tool that allows one to control the nanoscale morphology (**Figure 3**) of the titania films with unique properties [34, 35]. Typical porosity values of the mesoporous titania thin films are 66 ± 2%. The amphiphilic diblock copolymer polystyrene-b-poly(ethylene oxide) (PS-b-PEO) was used as a structure directing template. A typical synthesis involves dissolving of PS-b-PEO (20 mg) initially in a good solvent (toluene, 6.7 ml), followed by adding 1-butanol (bad solvent, 2.68 ml). This was performed to induce micelle formation, and the precursor titanium(IV) isopropoxide (TTIP) (73.8 μl) was selectively incorporated into the PEO phase. This is followed by 30 min stirring. To analyze the role of HCl on the final nanostructure formation, two solutions, one solution containing 121 μl of 6 M HCl (called WHCl), and other without HCl (called NHCl) were used in the spray solution. Further the solutions (WHCl and NHCl) were stirred for 20 h (500 rpm) at ambient conditions. The remote-controlled spray gun was used to spray the solution on a silicon substrate, at a pressure (N_2 carrier gas) of 1.5 bar. After this film was deposited, polymer template was removed by calcination at 550°C for 3 h. The films formed in presence of HCl were of anatase phase and had small and uniform pores, while those without HCl, were polydisperse. The presence of HCl is said to promote hydrolysis over condensation reactions.

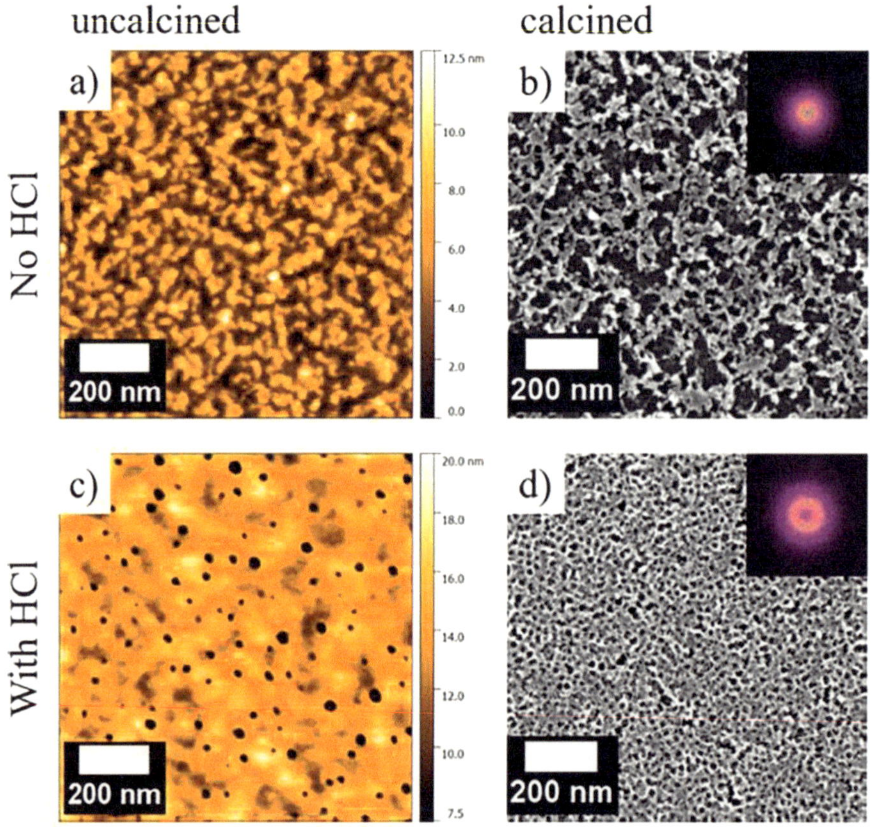

Figure 3.
(a, c) Surface topography (AFM) of titania thin films and (b, d) SEM images of titania thin films before and after polymer template removal. Reprinted with permission from Ref. [31]. © 2018 American Chemical Society.

2.1.2 Titania aerogels

Hydrolysis of precursors which are metal salts or metal alkoxides, followed by colloidal dispersion of solid precursors called sol. Further heat treatment results in complete polymerization and removal of solvent or water yields infinite network called solid gel phase. A wet gel is obtained by casting the sol into a mold. Further heating and drying removes the solvent to yield a highly porous, less-dense Aerogel.

Titania is a material of choice for catalytic applications, as it has high surface area and shows strong metal-support interactions. The titania samples prepared earlier by traditional methods, have reported low-to-moderate surface area. The most widely used Degussa P-25 TiO_2, was prepared by flame hydrolysis of titanium tetrachloride and is reported to have surface area of about 50 m^2/g [36]. The Glidden TiO_2 that was prepared by the hydrolysis of titanium isopropylate & subsequent firing at 500°C, has a surface area of about 90 m^2/g [37]. Some samples prepared by base hydrolysis of titanous chloride at different pH values, were dried in air at 110–120°C. These samples were reported to show surface area over 200 m^2/g [38].

Titania aerogels have been considered as promising photocatalysts. Ko et al. [39] have synthesized Titania aerogels by sol-gel process by controlled hydrolysis of methanolic solution of titanium-n-butoxide in presence of water and nitric acid. Subsequent removal of water was done by drying with supercritical CO_2. In a typical synthesis, TiO_2 gel was prepared by a simple sol-gel process using titanium-n-butoxide (TIB) as precursor in ethanol, deionized water, and hydrochloric acid mixture. TIB was dissolved in 40 ml alcohol in a dry glove box. To this, solution containing 10 ml alcohol, nitric acid and DI water. Concentrations of TIB, water and acid were kept at 0.625 mmol of TIB/ml of alcohol, 4 mol of water/mol of TIB, and 0.125 mol of nitric acid/mol of TIB, respectively. The solution was stirred to obtain a clear gel which was allowed to age for 2 h. Then it was extracted in an autoclave with supercritical CO_2 at flow rate of 24.6 L/h, a temperature of 70°C, and a pressure of 2.07 × 107 Pa (3000 psi).

The conventional solvent removal method of drying the aerogel collapses the porous network due to tension at the liquid-vapor interface. Water and nitric acid contents in the hydrolysis reactions were varied in the process to achieve titania aerogels (**Figure 4**) with maximum surface area. Thus obtained TiO_2 aerogel had a BET surface area exceeding 200 m^2/g, after calcination at 500°C for 2 h. This titania sample contains mesopores of 2–10 nm size and was of pure anatase form, which was shown by the Raman spectral bands at 441 cm^{-1}.

The commercial TiO_2 (Degussa P-25) is a mixture of anatase and rutile phases, as shown by their XRD & Raman spectra. The anatase form of titania is of more interest in catalysis applications than the rutile form, because of the higher surface area of anatase than the rutile (7 vs. 200 m^2/g). Compared to Degussa P-25, the titania aerogel prepared by the sol-gel process [40], had a surface area four times larger and was of pure anatase form. Tomkiewicz et al. [40] found a correlation of morphology of the TiO_2 aerogels with its catalytic activity. It was prepared by sol-gel process by dissolving titanium isopropoxide precursor in absolute ethanol and then mixing it with ethanol, DI water and nitric acid at concentrations ((Ti/ethanol/H_2O/HNO_3 = 1:20:3:0.08 ratio), followed by aging of the gel in alcohol for few days to weeks. The gel was then dried with CO_2 at its supercritical point (35°C and 1200 psi). This yielded aerogel with low density (0.5 g/cc) and high porosity (85%).

Dagan and Tomkiewicz [41] prepared titania aerogels using sol-gel process and supercritical drying (Ti/ethanol/H_2O/HNO_3 = 1:20:3:0.08 ratio). The aerogels had a surface area of 600 m^2/g and 85% porosity, compared to the Degussa P-25.

Figure 4.
SEM image of TiO_2 Aerogel. Reprinted with permission from Ref. [40]. © 1995 American Chemical Society.

Photodegradation of salicylic acid using titania aerogel photocatalyst under near-UV light (2 h) was found to be 10 times much faster than the Degussa sample.

2.1.3 Titania opals

The word opal refers to a gemstone which exists in nature. Chemically, it is hydrated amorphous silica with water content varying between 6% and 10%. As for the titania nanostructures, opal has come to mean the shape of the nanostructure. The ordered arrays of TiO_2 opals (**Figure 5**) were prepared using opal gel template-assisted sol-gel process under uniaxial compression at ambient temperature [42].

Typically, silica opal was used as template to synthesize polystyrene (PS) inverse opal. An aqueous HF (40%) solution was applied to remove the silica template. Monomer solutions containing dimethylacrylamide, acrylic acid, and methylene bis-acrylamide (wt% ratio: 1:1:0.02) was prepared in aqueous ethanol (4:7 w/w) with a 30 wt% limit on total monomer content. Ethanol is preferred here to enable the diffusion of monomer solution into the inverse opal PS. Then, 1 wt% of the initiator 2,2′-azobisisobutyronitrile (AIBN) was added to the monomer solution to initiate free radical polymerization at 60°C for 3 h. The inverse opal PS template was removed by Soxhlet extraction, to obtain Opal gel. The opal gels with different properties can be obtained by modifying the monomer solution, hole sizes, and stacking structures of the inverse opal template.

Then, opal gel template was placed into large quantity of tetrabutyl titanate (TBT) at ambient temperature for 24 h. Thus, the swollen opal gel was immersed into water-ethanol mixture (1:1 w/w) for 5 h. During this time, TiO_2 sol-gel process begins to complete the formation of opal structure of the gel. Subsequent calcination yields TiO_2 opal with distinctive spherical contours.

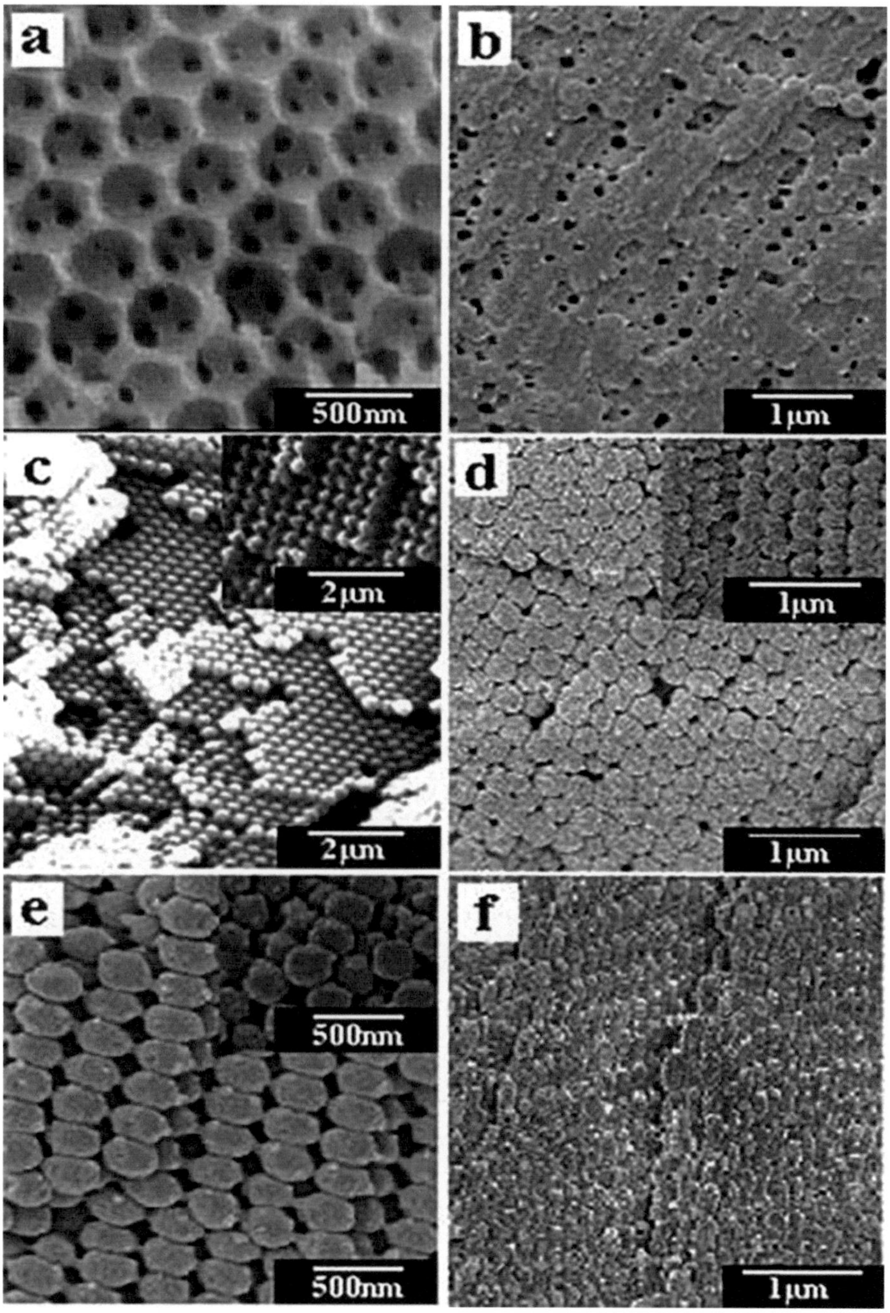

Figure 5.
SEM images of: (a) the inverse polystyrene opal. (b) The hydrogel opal after freeze-drying. (c) The gel/titania composite opal without compressing the opal gel template during the sol-gel process. (Inset) image of the sample after calcined at 450°C for 3 h. (d–f) (main panel) oblate titania opal materials after calcined at 450°C for 3 h, subject to compression degree R of (d) 20%, (e) 35%, and (f) 50%. Reprinted with permission from Ref. [42].

3. Modifications of TiO_2 nanomaterials

TiO_2 nanomaterials have bandgap greater than 3 eV and so they are transparent to visible light. TiO_2 nanomaterials have found applications such as photocatalysis, photovoltaics, Sensing, electrochromics, photochromics, UV protection, photo-induced water-splitting that are largely dependent on its optical absorption characteristics. TiO_2 nanomaterials absorb in UV (higher energy) region because of their wide bandgap, and this limits the performance of the nanomaterials. Improving the performance of the TiO_2 nanomaterials is to shift its absorption from UV to visible region, that is, the nanomaterials perform better by using less amount of energy. This can be done by: (i) Doping the nanomaterials with suitable metal ion which can narrow the electronic bandgap and alter its optical properties; (ii) adsorbing a colored inorganic/organic compound on to the nanomaterial, i.e., sensitization of TiO_2 nanomaterial can improve its visible absorption edge; (iii) coupling the electrons in the Conduction band of metal nanoparticle surface with those in the conduction band of TiO_2 nanomaterial in a metal–TiO_2 nanocomposite. By doping or sensitization, the visible-absorbing and more active Titania nanomaterials have been obtained as evidenced by their utility in environment (photocatalysis, sensing) and energy (photovoltaics, water-splitting, photo-/ electro-chromics, H_2 storage) fields for a sustainable development.

3.1 Synthesis of doped TiO_2 nanomaterials

Some organic compounds such as nitro-stilbene derivatives were found to show non-linear optical (NLO) activity. Sol-gel process is a wet chemical synthetic route that allows for the incorporation of optically active organic molecule into the inorganic metal-oxide glass matrix to obtain a doped gel with specific optical properties. The inorganic metal-oxide glass was found suitable for stabilizing the NLO materials because of higher thermal stability of the metal-oxide glass, compared to polymer [43]. The sol-gel process was used to fabricate titania films doped with NLO material, for use in electro-optic devices [44]. In a typical sol-gel process, precursor solution was prepared by mixing HCl/H_2O with isopropanol (PrOH) solution of tetraisopropoxytitanate (TPOT) containing the NLO materials with vigorous stirring at ambient temperature, at a molar ratio of 1:1:0.5:1.88 for $TPOT:PrOH:HCl:H_2O$. Final solution was used to spin-coat to obtain 1-micron-thick titania film on Indium Tin Oxide (ITO) glass substrate.

3.2 Synthesis of metal-doped titania nanomaterials

The TiO_2 nanoparticles were doped with 21 different metal ions by means of sol-gel process and this made a significant impact on photoreactivity, charge-recombination rate and interfacial electron-transfer rates, in a TiO_2-nanomaterial-based photovoltaic device, where photon-to-current efficiency is largely dependent on these factors [45].

Li et al. [46] found that La^{3+}-doping of TiO_2 by sol-gel process, could impart thermal stability, prevent phase transformation, reduce nanoparticle size, and increase Ti^{3+}-content on the surface.

The dopant Nd^{3+} ion (1.5 at%) in the TiO_2 nanoparticle, introduces energy level into the bandgap of the nanomaterial, to be the new LUMO. Thus, the dopant brings down the bandgap by 0.55 eV [47]. This new LUMO level brings down the energy of the bandgap, and this can shift the absorption onset of TiO_2-nanomaterials from UV to visible region, thereby altering its optical properties [48].

This bandgap narrowing results in a red-shift in the absorption of metal-doped TiO_2. With increase in atomic number of the metal-ion dopant, an energy level that is formed shifts the localized level to lower energy [49].

Pt-doped titania thin films were synthesized using sol-gel process, followed by dc magnetron sputtering for the Pt film deposition. Substrate sapphire wafer was cleaned in ultrasonication bath in sequence, in acetone, isopropanol and deionized water for 15 min and subsequently dried at 90°C for 5 min. The titania thin films were first prepared by sol-gel process by mixing 3.8 ml of ethanol with 0.7 g of Triton-X-100 under ambient conditions for 3 min. To this mixture, 0.68 ml of acetic acid and 0.36 ml of titanium (IV) isopropoxide were added. Using this sol, spin-coating process was repeated three times with a speed of 3000 rpm and short intermediate heating at 550°C. The titania thin films were then annealed at temperatures from 600 to 1000°C at the rate of 100°C/15 min. Then, a Pt thin film (20 nm thick) was deposited using dc magnetron sputtering. Ti-alkoxide was formed in situ by the esterification of alcohol by acid, and the alkoxide was hydrolyzed in presence of the non-ionic surfactant Triton X-100 to organize the material structure and well-defined nanophases. Annealing of the thin films at 600–800°C yielded anatase phase with smaller grains (15–28 nm), while the higher temperature (900–1000°C) annealing gave rise to rutile phase with larger grains (100–130 nm), high surface roughness and reduced bandgap energy (2.8 eV, compared to anatase 3.4 eV). Higher temperature annealed thin films showed higher sensitivity (10^3–10^4) to hydrogen gas that has been attributed to increased roughness and a greater number of adsorption sites [50].

3.3 Synthesis of TiO_2-nanoparticle-shelled microspheres

New, 'Open Mouth', Hollow TiO_2-nanoparticle-shelled (OMHTNPS) light-driven microcleaners were obtained via low-cost, high-throughput, facile sol-gel method, with the subsequent removal of carbon microspheres by a simple sintering process [51]. The shells of the prepared OMHTNPS microcleaners mainly contain 20-nm anatase TiO_2 nanoparticle. The carbon microspheres (CMSs) and TiO_2-coated CMSs (TiO_2@CMS) were prepared as reported [52]. These OMHTNPS showed 98% efficiency in the photodegradation of the Rhodamine B dye.

Synthesis of Janus micromotors: TiO_2 microspheres were prepared by the solvent extraction/evaporation method using tetrabutyl titanate as a precursor. Briefly, 1.0 ml of tetrabutyl titanate was dissolved in 40.0 ml of ethanol and incubated at room temperature for 3 h; then, TiO_2 microspheres were collected by centrifugation at 7000 rpm for 5 min and washed repeatedly with ethanol and ultrapure water (18.2 MΩ cm), three times each, then dried in air at room temperature. TiO_2 (anatase) microsphere is obtained after annealing for 2 h at 400°C. The X-ray diffraction (XRD) pattern reveals that the TiO_2 microspheres have a good anatase phase. For the TiO_2–Au light-driven Janus micromotor, TiO_2 microspheres (1.0 μm mean diameter) are used as the base particles. TiO_2 particles (10.0 μg) were first dispersed in 150.0 μl of ethanol. The sample was then spread onto glass slides and dried uniformly to form particle monolayers. The particles were sputter coated with a thin gold and nickel layer using a Sputter Coater for 3 cycles with 60 s per cycle. The metal layer thickness was found to be 40 nm, as measured by the Profilometer. For the TiO_2–Ni–Au magnetic Janus motors, TiO_2 particle monolayers were prepared as in the method above. A 40 nm layer of Au followed by a 10 nm layer of Ni were sequentially deposited on half of the particles by Sputter Coater. The TiO_2 microspheres were coated with Al_2O_3 layer using ultrahigh Vacuum Magnetron Sputter Coater. The micromotors (**Figure 6**) were subsequently released from the glass slides via pipet pumping and dispersed into double distilled water. The polystyrene–Au Janus microsphere as a control was fabricated with the same method using polystyrene microsphere. These light-driven, precisely controllable, and highly efficient TiO_2-based photocatalytic Janus micromotors offer possibilities in designing such light-driven nanomachines for a range of applications from nanofabrication [53] to environmental remediation [54].

3.4 Titania-hybrid photonic crystals

Titania-hybrid photonic crystals are high dielectric lattices that find applications in light control for waveguiding and lighting devices, photocatalysis, photovoltaics, and sensing. Sol-gel processing used for their fabrication allows for the alternated spin-casting of high and low refractive index polymer solutions or the sol of titania particles and subsequent sintering. This solution-processing method has attracted interest owing to simpler structures, ease of fabrication, efficient scale-up, low-cost

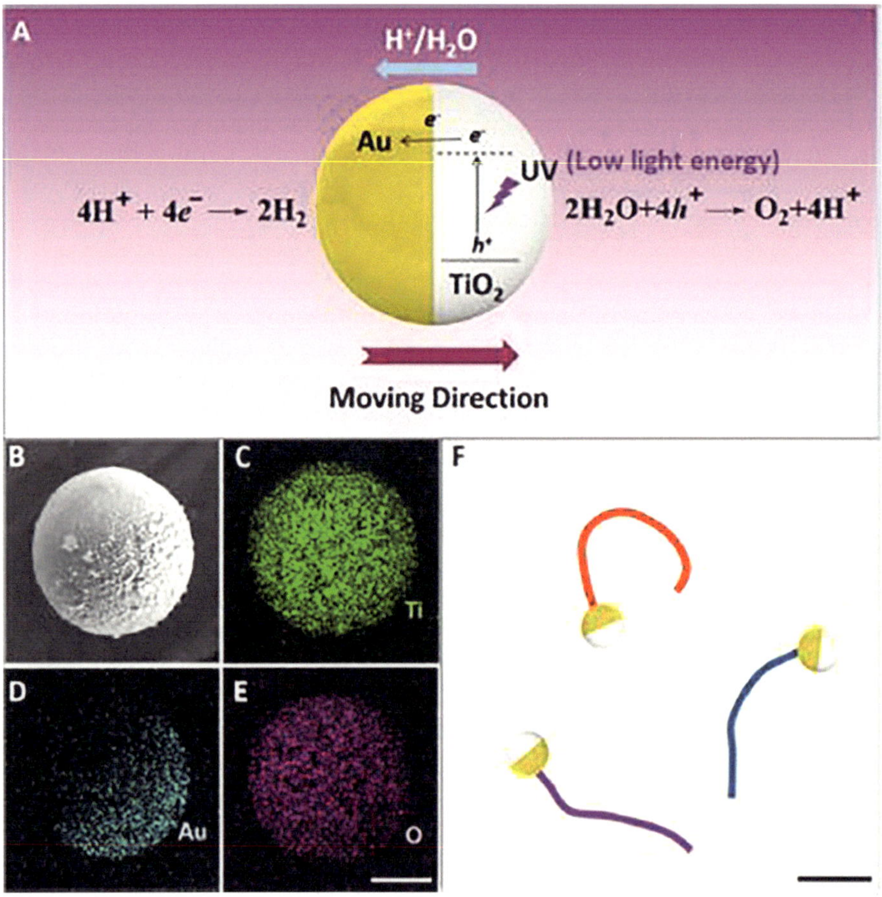

Figure 6.
Catalytic scheme, SEM and EDX images of Au–TiO_2 micromotor. (A) Schematic of Catalytic TiO_2–Au Janus Micromotors powered by UV Light in water. (B) SEM image of a spherical TiO_2–Au micromotor. (C—E) The corresponding EDX images for Ti, Au, O, respectively. Scale bar, 0.5 μm. (F) Tracking lines illustrating the distances traveled by three micromotors in pure water over 1 s. Scale bar, 10 μm. Reprinted with permission from Ref. [53]. © 2016 American Chemical Society.

processing, and offers the product, flexibility, and permeability [55]. First Titania sol was prepared by adding 10 ml of polyacrylic acid (PAA) in butanol and catalytic amount of HCl (100 μl) to 10 ml of titanium butoxide precursor solution. The organic & inorganic components in the deposited films can be varied using different concentrations of titania precursor and PAA in the initial solution. The resultant solution is hydrolyzed by stirring at room temperature for 2 h, when a transparent sol is formed. Thin films and distributed Bragg reflectors (DBRs) were obtained by spin-coating of the sol and of the polymer solutions, at a speed of 5400–12,000 rotations per minute. The Ti-Hybrid was heated subsequently at 80°C, while the Si-Hybrid heated at 300°C. Multilayers were grown by spin-coating of alternated high (Ti-Hy) and low (Si-Hy, PMMA) refractive index media with alternating layer of PMMA, to form a DBR.

3.5 Template-assisted sol-gel syntheses

3.5.1 AAM-template-assisted synthesis of TiO_2 nanotubes

The TiO_2 nanotubes (**Figure 7**) were synthesized by sol-gel process in combination with Anodic Alumina Membrane (AAM) used as a template [56].

In a typical synthesis, a thin layer of TiO_2 sol is drawn into the pores of the AAM under vacuum. The TiO_2 sol was prepared by sol-gel process using Titanium tetraisopropoxide (TTIP) as Ti precursor. The TTIP solution was prepared by mixing TTIP with isopropanol and 2,4-pentanedione. After dipping the AAM template into this solution, the entire solution was drawn through the pores of AAM under vacuum. The Titania nanotubes were obtained after dissolving the membrane in 6 M NaOH solution for several minutes [57].

3.5.2 ZnO-nanorod as template

Zinc oxide nanorod array on a glass substrate was used as a template to fabricate TiO_2 nanotubes by sol-gel method [58]. By this method, TiO_2 sol was prepared first and the ZnO nanorod template was dip-coated by immersing in the sol and taken out at a slow speed, dried at 100°C for 10 min, further heated in air at 550°C for 1 h, to obtain ZnO/

Figure 7.
SEM image of titania nanotubes using AAM template. Reprinted with permission from Ref. [56]. © 2005 American Chemical Society.

TiO_2 nanorod arrays. The template was removed by immersing the ZnO/TiO_2 nanorod arrays into dilute HCl solution. The TiO_2 nanotubes have a length of 1.5 micron, with inner diameter of 100–120 nm that is characteristic of the ZnO nanorod template. To get a well-aligned TiO_2 nanotube array, an optional dip-coating cycle (2–3 cycles) can be adopted.

3.5.3 Template-based sol-gel electrophoretic deposition

Electrophoresis is a type of motion of charged particles in a colloidal system or a sol, in response to the application of external electric field. When a charged particle is in motion, the solvent part tightly bound to the particle will move with it, whereas the counter-ions diffuse in the opposite direction. The electrophoretic deposition technique uses such an oriented motion of charged particles to grow films or monoliths by enriching the solid particles from a sol (prepared by sol-gel process) onto the surface of an electrode.

Limmer et al. [59] combined sol-gel synthesis and electrophoretic deposition in the growth of nanorods of various oxides including complex oxides such as barium titanate. Similar procedure was used to grow nanorods (**Figure 8**) of TiO_2. In a typical

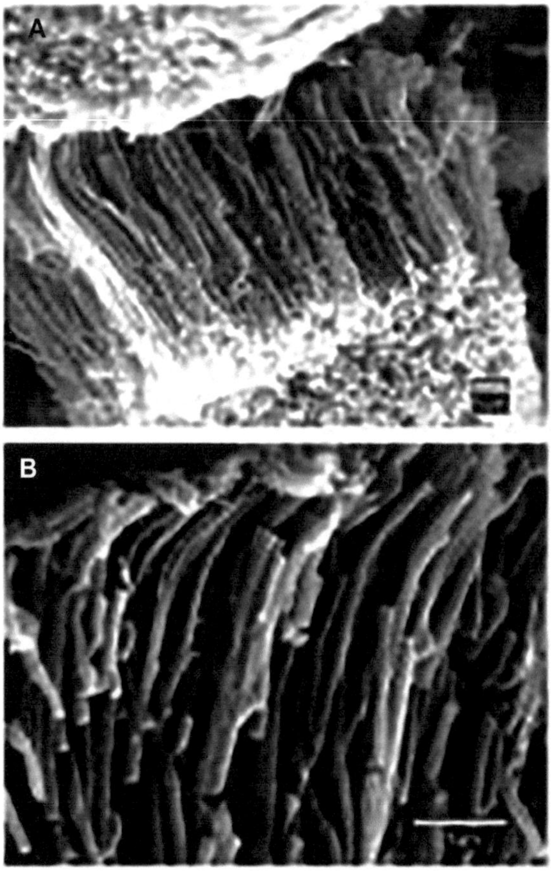

Figure 8.
SEM images of TiO_2 nanorods grown in a PC membrane with 200 nm diameter pores by sol—gel electrophoretic deposition. (A) Lower magnification image, showing that the rods are aligned and grown over a large area. Scale bar, 1 μm. (B) Higher magnification image of the nanorods. Examination of broken rods seen here shows that they are solid and dense. Scale bar, 1 μm. Reprinted with permission from Ref. [60]. © 2002 John Wiley and Sons.

process, conventional sol-gel processing was used to prepare TiO_2 sols. By maintaining appropriate pH, electrostatically stabilized, nanoparticles dispersed uniformly in solvent were obtained with desired stoichiometric composition [60]. When an external electric field is applied, these nanoparticles move and deposit on the cathode or anode, depending on the zeta potential (surface charge) of the nanoparticles. Using radiation track-etched polycarbonate membranes with an electric field of 1.5 V/cm, nanowires with diameters of 40–175 nm and a length of 10 microns corresponding to the thickness of the membrane. By this method, many complex oxides ($BaTiO_3$, $Sr_2Nb_2O_7$) and inorganic-organic hybrids with desired composition, have been synthesized [61].

3.5.4 Template-based electrochemical sol-gel deposition

Single crystal TiO_2 nanowires were synthesized by this sol-gel deposition method [62]. The electrolyte solution was prepared according to the work of Natarajan and Nogami [63]. First, Titanium powder was dissolved in a mixture of H_2O_2 and ammonia solution, then the excess H_2O_2 and ammonia were decomposed by heating the solution on a hot plate and, consequently, a yellow-colored gel was obtained. By dissolving the yellow gel in 4 M H_2SO_4, a red-colored solution formed and the red-colored solution was used as the stock solution for further electrodeposition. A certain amount of KNO_3 was added to the stock solution (about 145 mM), and the pH adjusted to 2–3 by using ammonia solution. The resultant solution was used as electrolyte in the electrodeposition process.

Electrodeposition was carried out at room temperature (20–25°C) using a three-electrode potentiostatic system which comprises SCE (reference) electrode, 2 cm × 1.5 cm Pt plate as counter-electrode and a small piece of AAO template with Au substrate as working electrode. The porous side of the working electrode was exposed to the electrolyte. The templates with pore diameters of 50, 22, and 20 nm were used in the fabrication. The deposition was carried out under potentiostatic conditions at −0.9 to 1.2 V. As a result, nanorods of amorphous TiO_2 gel formed. Subsequent heat treatment at 450°C for 24 h in air, yielded nanowires of single crystal TiO_2 with anatase (**Figure 9**) structure (diameters of 10, 20, and 40 nm and lengths of 2–10 μm).

Figure 9.
SEM of single-crystal TiO_2 nanowires. Reprinted with permission from Ref. [62]. © 2002 American Chemical Society.

The electrophoretic sol-gel method failed to synthesize nanorods of diameter less than 50 nm. Compared to the electrophoretic sol-gel process, the electrochemical sol-gel deposition technique has advantages: (i) It could readily achieve nanowires of diameter less than 20 nm, as templates with very small pores (<20 nm) can be used; (ii) lengths of nanowires can be controlled by varying deposition time and potential of the working electrode; and (iii) high local pH at the AAO pores causes hydrolysis, gelation, and aging processes. This forms a more compact gel structure with higher packing density, less shrinkage and less cracking.

3.6 Syntheses of TiO_2 nanoparticles

Titania nanoparticles of different sizes and shapes were obtained by sol-gel process involving the precursor TTIP under appropriate reaction conditions. Sugimoto et al. [64–68] developed the synthesis process by series of studies. The synthesis process consists of preparing a stock solution of titanium source (0.5 M Ti), by mixing the precursor TTIP with triethanolamine (TEOA) {[TTIP]/[TEOA] = 1:2} and water. The stock solution is diluted with shape controller solution (Amine) and then aged at 100°C for 1 day and at 140°C for 3 days. The pH of the solution is varied from 0.6 to 12, by adding $HClO_4$ or NaOH. With increase in pH, yield of the nanoparticles decreases to 9% (pH 12). This suggests that varying the pH had significantly decreased the nucleation rate of Anatase TiO_2, by reducing the concentration of the precursor. Amines used in the process include ethylene diamine, diethylene triamine, triethylene tetraamine, trimethylene diamine, TEOA. These amines function as shape controller as well as surfactants.

3.7 Inorganic sensitization

3.7.1 Sensitization by narrow bandgap semiconductors

Narrow bandgap semiconductors have been used as sensitizers to increase the visible absorption edge of the titania nanomaterials that have wide bandgap. This shifts the optical absorption to visible region so that these nanomaterials generate photocurrent with less energy. The inorganic semiconductor-sensitized TiO_2 nanostructures have been prepared usually by the sol-gel process [69–73].

Semiconductor PbS-sensitized TiO_2 nanocrystalline system has enabled quicker injection of photogenerated electrons from the PbS into the TiO_2 nanomaterial and generated strong photocurrent using visible light [74]. The nanocrystalline TiO_2 films were prepared using standard sol-gel techniques. First colloidal TiO_2 solutions were prepared from titanium isopropoxide (30 ml) and isopropanol (10 ml) in water (500 ml). Nitric acid was added to adjust the pH to 1. The organic components were evaporated by boiling the solution for 12 h and crystallization of TiO_2 particles resulted. The colloidal solution was then spin coated on glass substrates, provided with evaporated Cr contacts in planar geometry for the conductivity measurements. The freshly deposited films were heated for 5 min at 450°C. Coating and drying were repeated several times to get a film of desired thickness of 1 mm. At the end of this process, the samples were baked at 450°C for 30 min. The films consist of anatase crystallites (40–60 nm diameter), and are structurally stable up to 650°C. The internal surface area is 400 times larger than the projected area. The conductivity of the films is typically below $\sim 10^{-9}$ $(\Omega\ cm)^{-1}$ at room temperature. PbS clusters adsorbed on the internal surface of the TiO_2 clusters were then prepared by dipping the TiO_2 films into concentrated lead acetate solution and subsequently precipitating the adsorbed Pb^{2+} with a solution containing sodium sulfide.

The PbS colloidal particles of about 30 Å diameter were obtained. Larger particles were formed by repeating the dip-coating process several times. The residual water in the films was removed by heating to 200°C at reduced pressure.

Figure 10 shows the photo-action spectra for the bare TiO_2 and the TiO_2–PbS films obtained after 1, 2, and 5 coatings of PbS. In the bare TiO_2 film, the onset of photoconduction is found to occur at 380 nm, corresponding to the 3.2 eV band gap of anatase (bulk) TiO_2 [75]. In the films coated with PbS, the TiO_2 response vanishes due to the high absorption of the PbS clusters in this wavelength region. Instead, a broad band due to the PbS emerges in the visible region with a maximum around 500 nm. Optical transmission spectra of the three TiO_2–PbS films (1, 2, and 5 coatings) were compared which showed a red-shift of the absorption, with increase in the number of PbS coatings, as shown by the absorption edge at 1.6 eV, 1.37 eV and 1.24 eV, respectively. This is indicative of efficient sensitization of TiO_2 by PbS. This has been related to the average PbS cluster sizes of 28, 35 and 40 Å. The clusters of size below 25 Å have been found to be more efficient sensitizers that gave rise to better photoconduction response with increase in light intensity.

Fitzmaurice et al. [76] found rapid electron injection into the TiO_2 electrode was possible with AgI-sensitization, as evidenced by the enhanced lifetime (>100 μs) of electron–hole pairs.

Typical synthesis involves the preparation of TiO_2 sol and AgI sol separately and mixing them later. TiO_2 sol was made at acidic pH (3.3) by the hydrolysis of $TiCl_4$. This TiO_2 sol was made alkaline (pH 11.4) by rapidly mixing it with required amount of sodium hydroxide. Aqueous AgI sol was made by rapid mixing of appropriate concentrations of $AgNO_3$ and KI solutions in presence of PVA stabilizer of concentration 0.002–0.2%.

The AgI–TiO_2 sandwich colloids were made by precipitation of AgI on the surface of TiO_2 particles. Silver nitrate solution (5×10^{-4} M) was mixed with alkaline solution of TiO_2 (1 g/L). After 20 s, the resultant solution was rapidly mixed with equivalent amount of KI. The solution was allowed to age for 12 min prior to use.

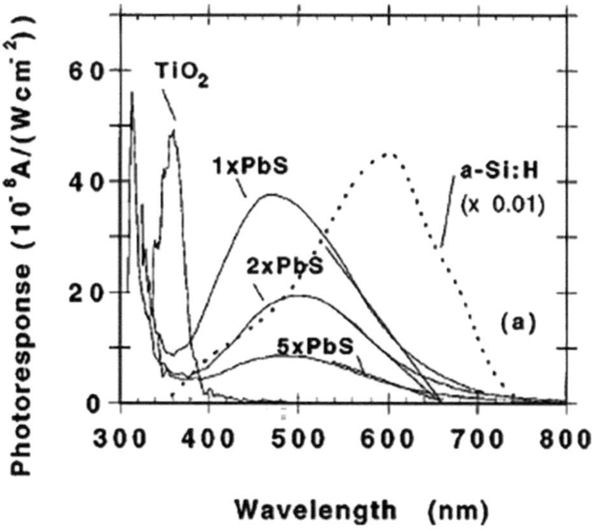

Figure 10.
Photo-action spectra of TiO_2 film with multiple coatings of PbS nanocrystallites. Reprinted from Ref. [75], with permission of AIP Publishing.

Wide bandgap semiconductor particles (TiO_2) with low-lying conduction band are combined with the narrow bandgap semiconductor (AgI, CdS) particles with high-lying conduction band. Upon illumination, electrons are transferred from AgI into the conduction band of Titania, while the holes remain with narrow bandgap semiconductor. This results in efficient electron transfer to the wide bandgap particles and minimizing the charge-recombination. This sensitization can potentially enhance the performance of AgI-sensitized TiO_2 nanostructures in photovoltaic cells, photocatalysis, new generation display monitors, non-linear optics [77].

Vogel et al. [78] extensively studied the sensitization of TiO_2 by different semiconductors CdS, Ag_2S, Bi_2S_3, Sb_2S_3 and found that efficient charge separation and photostability of TiO_2 could be achieved by surface modification of the titania nanostructures by such semiconductors. The relative positions of energy levels at the CdS–TiO_2 interface could be optimized for efficient charge separation, using the size quantization effect. **Figure 11** shows the plot of photocurrent quantum yields (IPCE) for four differently treated TiO_2–PbS electrodes versus the illumination time (λ = 460 nm, 8 mW cm^{-2}). Curve 1 refers to a decrease of high initial IPCE (65%) value in first few minutes of illumination for opaque microporous TiO_2 electrode. Curve 2 shows a clear improvement in photostability of the transparent PbS-sensitized TiO_2 electrode. Additional coating of CdS resulted in the increased IPCE value (74%). Further illumination for 4 days yielded the curve 4 and the photostability of the electrode is strongly enhanced.

The energy levels of sensitizer–substrate junction can be tailored by varying the energy levels of the sensitizer taking advantage of the size quantization effect and keeping the energy level of the substrate constant. As the particle size of sensitizer approaches that of the bulk, its lowest edge of the conduction band lies below that

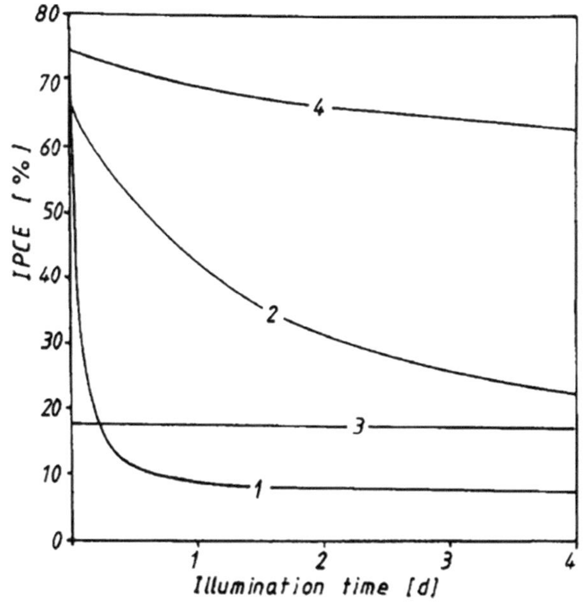

Figure 11.
Photocurrent quantum yields for differently treated PbS–TiO_2 electrodes as a function of the illumination time with X = 460 nm and p = 8 mW cm^{-2}. Curve 1: one coating with PbS on a microporous TiO_2 substrate. Curves 2–4: one coating with PbS on a nanoporous TiO_2 substrate. Curve 2: as prepared. Curve 3: after deposition of a thin TiO_2 layer. Curve 4: after one additional coating with CdS. Reprinted with permission from Ref. [78]. © 1994 American Chemical Society.

of the TiO_2 and electron transfer from the sensitizer to TiO_2 cannot occur. With CdS semiconductor nanoparticles as sensitizer, electron transfer from the excited CdS into the Titania electrode occurred only when the particle size of the CdS was sufficiently larger than 2 nm, suggesting the role of quantum size effects in the charge-transfer process. For a wide bandgap substrate like TiO_2, optimum particle size of the CdS for high photocurrent quantum yield has been found to be 4–5 nm.

The sensitization of the TiO_2 films with CdSe semiconductor nanoparticles shifted the absorption to visible region of the electromagnetic spectrum. Upon irradiation in visible region, the CdSe–TiO_2 composite photoanode in a photoelectrochemical cell showed IPCE of 12% due to rapid electron injection from the CdSe into the TiO_2 [79].

3.7.2 Sensitization by metal nanoparticles

Nanoporous TiO_2 films were loaded with Au nanoparticles and the Au nanoparticles were photoexcited due to plasmon resonance. Then charge separation occurred by the transfer of electrons from the Au nanoparticles into the TiO_2 film and by the electron transfer from donor in solution to the Au nanoparticles [80]. Similar loading of Au/Ag nanoparticles into the TiO_2 film was potentially useful in applications such as Photovoltaics, Plasmon sensors, photocatalysis [81]. Upon UV-light illumination, the TiO_2 nanorods sensitized with Au/Ag nanoparticles were found to sustain higher degree of conduction band electrons, compared to pure TiO_2 [82].

3.7.3 Sol-gel deposition of 2D array of Au/Ge nanodots on patterned TiO_2

Self-assembled Inorganic NanoPatterns (INPs) on crystalline silicon wafers as templated surfaces have been explored for the formation of Au and Ge nanoparticles. The substrates were prepared by sol-gel liquid deposition and evaporation induced self-assembly (EISA) of a hybrid solution composed of block copolymer micelles and TiO_2 inorganic metal oxide precursor. This resulted in a single layer of hexagonally arranged micelles surrounded by the inorganic precursors. After condensation and block copolymer decomposition by heat treatment, the final thin metal oxide layer bears uniform nanoperforations with controlled spacing (100–510 nm) and height (5–15 nm) according to the length of the block copolymer in use. Such arrays of self-organized metal nanodots on the TiO_2 nanopatterns have been studied for their optoelectronic applications [83]. In a typical synthesis, Sol-gel initial solutions are composed of $TiCl_4$:EtOH:H_2O:$PB_{12.5}$-b-PEO_{15} (molar ratio = 1:40:7:1.5 × 10^{-3}). $PB_{12.5}$-b-PEO_{15} refers to polybutadiene-block-poly(ethylene oxide) with blocks of 12,500 and 15,000 g mol^{-1}. In the case of $PB_{5.5}$-b-PEO_{30}, the molar ratio is 10^{-3}. The solution is divided in two parts: in part A, PB-b-PEO is dissolved in 2/3 of the ethanol and water; part B contains $TiCl_4$ and the remaining ethanol. The solutions are aged for 2 h at 70°C, and then part A is slowly cooled to room temperature in ~30 min. Finally, both parts are mixed before use. Films are deposited on cleaned silicon wafer by dip coating at a temperature of 40°C and a relative humidity below 20%, using a withdrawal speed in the range of 1–3 mm s^{-1} to obtain a film thickness of <10 nm, corresponding to a monolayer of INPs (**Figure 12**). Additional SEM characterization can be performed to assess that only a monolayer is deposited and to modify the withdrawal speed if not. The resulting film is then annealed at 450°C for 30 min. Substrates, previously dip-coated to obtain self-assembled perforations, are immersed into a diluted hydrofluoric acid (HF) solution of 1.17 mol^{-1} for 20 s to remove the native silicon oxide at the bottom of the perforations and reveal the silicon surface

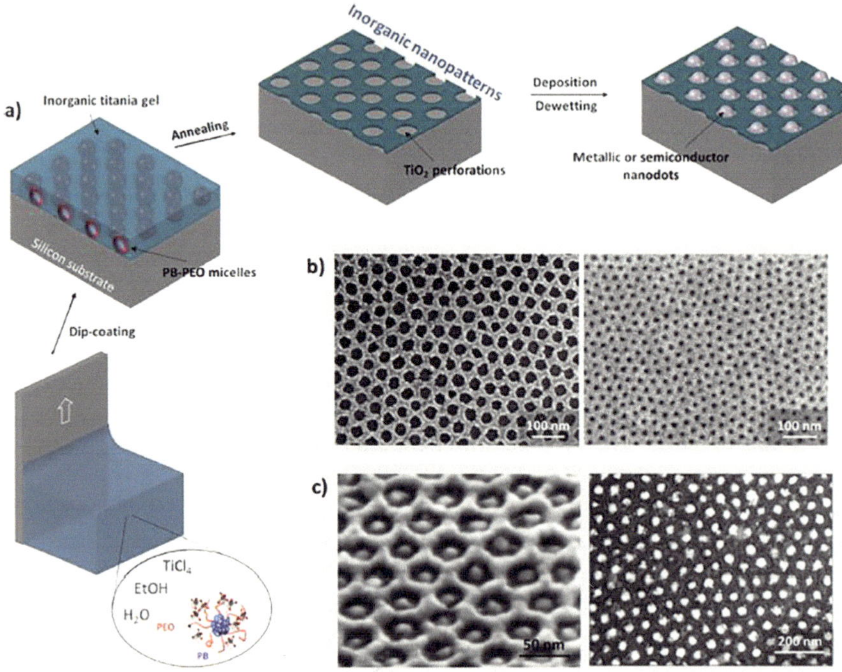

Figure 12.
(a) Scheme depicting the process to obtain organized metallic nanodots. A monolayer of micelles embedded in a titania gel is first deposited on a silicon substrate. After annealing, TiO_2 INPs are formed revealing the bare silicon. Under the appropriate conditions, a single nanodot per perforation is obtained. (b) SEM images of the TiO_2 INPs network after annealing for (left) large perforations of 20 nm ($PB_{12.5}$-b-PEO_{15}), (right) small perforations of 12 nm ($PB_{5.5}$-b-PEO_{30}). (c) SEM images of nanodots hexagonally arranged in TiO_2 INPs: (left) nanocrystalline Ge nanodots and (right) Au nanodots. Reprinted with permission from Ref. [83]. © 2019 American Chemical Society.

without damaging the INPs. Open perforations of 28 ± 4 nm in diameter are obtained with accessibility of the substrate surface. Immediately after HF treatment, the INP substrates are placed under vacuum. Gold is deposited by sputtering at room temperature ($P = 4 \times 10^{-6}$ mbar).

Using block copolymer–micelles-assisted sol-gel deposition of TiO_2 on Si and thermal annealing, the substrates with INPs featuring hexagonally positioned perforations homogeneously sized and spaced, were prepared. These templates are used to selectively form individual nanodots in each perforation featuring typical size of 28 ± 5 nm for the Au nanodots.

4. Performance of TiO_2 nanomaterials

H_2 storage: TiO_2 nanotubes were found to store H_2 gas up to ~2 wt% at room temperature and a pressure of 6 MPa (at atomic ratio of H/TiO_2 of 1.6), compared to a much lower hydrogen concentration of 0.8 wt% for the bulk TiO_2. Of this 2 wt% of the adsorbed hydrogen, only 75% could be released at lower pressure, while the remaining 25% tend to be retained owing to chemisorption. Only a part (13%) of the chemisorbed hydrogen was completely released from the Nanotubes after heating at

70°C [84]. Bavykin et al. [85] found that the H_2 gas was adsorbed between the layers of multilayered walls of the Titania nanotubes in the temperature range −195°C to −200°C at 0–6 bar pressure.

Electrode in DSSC: DSSC with electrodes made of TiO_2 nanotubes (10-nm diameter & 30–300 nm long) showed an efficiency of 4.88% and short-circuit current density more than twice that showed by the device made of Degussa P-25 TiO_2 nanoparticle electrodes, under AM 1.5 illumination [86].

Ohsaki et al. [87] found that better efficiency of solar cells fabricated using TiO_2 nanotube electrodes was due to increase in electron density in TiO_2 nanotube electrodes, compared to the bulk TiO_2 (P-25) electrodes.

Grimes et al. [88] fabricated DSSC with TiO_2 nanotubes (46-nm pore diameter, 17-nm wall thickness, 360-nm long) showed a photocurrent efficiency of 2.9%, which was attributed to superior electron lifetimes and electron percolation, compared to TiO_2 nanoparticle system.

Water-splitting: Br^-/Cl^- doped nanocrystalline TiO_2 electrodes were reported to shift the absorption edge to the visible region and showed better efficiency of water splitting than pure TiO_2 [89]. Nickel-doped TiO_2 photocatalyst was found to generate hydrogen gas at nearly 125.6 l mol/h compared to 81.2 l mol/h for pure P-25 TiO_2 [90].

Yang et al. [91] found that TiO_2 nanotubes treated with H_2SO_4 solutions showed photocatalytic activity on degradation of acid orange II in the following order: TiO_2 nanotubes treated with 1.0 mol/L H_2SO_4 solution > TiO_2 nanotubes treated with 0.2 mol/L H_2SO_4 solution > untreated TiO_2 nanotubes > TiO_2 nanoparticles, since TiO_2 nanotubes treated with H_2SO_4 were composed of smaller particles and had higher specific surface areas.

Electrochromic displays/windows: Electrochromism is the ability of a material to change color upon oxidation or reduction. The TiO_2 nanomaterials have been widely investigated for applications in electrochromic windows and displays. Electrochromic windows will darken upon application of a small voltage, while it will become transparent to visible light/solar light on reversing the voltage. A smart window can regulate the entry of light/energy through it in such a way that the need for air-conditioning the room decreases. Nanocrystalline structure of the TiO_2 film makes possible 100–1000-fold amplification compared to a flat TiO_2 surface. An electrochromophore molecule (adsorbed on to the nanocrystalline Titania electrode) switches color on applying a small voltage. High conductivity of nanocrystalline nature of the electrode, fast electron exchange with the electrochromophore, optical amplification by the porous structure and fast charge compensation by the ions in the contacting liquid, make nanocrystalline TiO_2 electrodes, highly attractive components of the electrochromic devices. These electrodes can be fabricated using sol-gel process followed by spin-coating to obtain a film of desired thickness [92, 93].

4.1 Conclusions

There have been continuous research efforts on the syntheses of TiO_2 nanomaterials in the past decades, owing to its attractive properties found critical to a wide range of applications such as photovoltaics, photoelectrochemical cells, photocatalysis, environmental/wastewater remediation, photo−/electro-chromics, opto-electronics, NL optics, flexible electronics, H_2 storage, and gas sensors, to name a few. There has been continuous research on the syntheses and modifications of similar non-magnetic metal oxide nanostructures [94]. The progress in synthesizing the technologically important TiO_2 with newer nanostructures and better properties, could not have been possible without the underlying research efforts in instrumentation as well. The sol-gel processing that was used earlier for the syntheses of metal oxide nanoparticles, has progressed to developing the TiO_2 nanomaterials with different morphologies such as microspheres, aerogels, opals, nanotubes, nanorods and nanowires. This progress was made possible by the sol-gel process with assistance of template, surfactants, micelles, NLO-active material, spin-coating, dip-coating, electrophoretic deposition, polystyrene, and diblock polymer. Further, sensitization of TiO_2 nanomaterials was made possible in sol-gel process using metal ion dopants, metal nanoparticles, narrow bandgap semiconductors, and organic dyes, depending on the type of sensitization required.

There are some practical concerns/challenges such as precise control of deposition at single atomic level, and growing best quality films, in fabricating metal-oxide thin films by the sol-gel process. There is a possibility of losing the porosity of the films during high-temperature sintering. But the porosity of the nanostructure is important for applications such as Catalysis, sensors for organic/biocomponent, electrodes in solar cells. Syntheses of crystalline phase of complex metal oxides, without the high-temperature sintering, needs to be addressed. Template-assisted sol-gel processing for the nanorod/nanotubes requires complete filling of the template/pores by sol and enrichment of solid inside the pores. Difficulty in ensuring the complete filling of the template pores needs to be addressed. There is a steady and continuous progress in the research on TiO_2 nanomaterials which will continue to impact the research on energy and environmental remediation fields. This continuing research on the titania nanostructures may possibly shed light on the synthetic process modifications needed to address these concerns and issues, without resorting to expensive instrumentation.

Acknowledgements

I thank my family for their immense support during the preparation of the chapter, and Lord Almighty for giving me the inner strength and clarity in this endeavor. Last but not the least, I thank our esteemed publisher, M/s IntechOpen Limited for offering me this authorship/opportunity and the continuous support.

Additional information

ORCID ID: 0009-0008-8249-6412.

Author details

Srinivasa Raghavan
Ramakrishna Mission Vivekananda College, Chennai, India

*Address all correspondence to: srirag11@gmail.com

References

[1] Pfaff G, Reynders P. Angle-dependent Optical Effects Deriving from Submicron Structures of Films and Pigments. Chemical Reviews. 1999;**99**:1963

[2] Salvador A, Pascual-Marti MC, Adell JR, Requeni A, March JG. Analytical Methodologies for Atomic Spectrometric determination of Metallic oxides in UV Sunscreen Creams. Journal of Pharmaceutical and Biomedical Analysis. 2000;**22**:301

[3] Braun JH, Baidins A, Marganski RE. TiO_2 Pigment Technology: A Review. Progress in Organic Coating. 1992;**20**:105

[4] Yuan SA, Chen WH, Hu SS. Fabrication of TiO_2 Nanoparticles/Surfactant Polymer Complex Film on Glassy Carbon Electrode and its Application to Sensing Trace Dopamine. Materials Science and Engineering: C. 2005;**25**:479

[5] Fujishima A, Honda K. Electrochemical Photolysis of Water at a Semiconductor Electrode. Nature. 1972;**37**:238

[6] Fujishima A, Rao TN, Tryk DA. Titanium Dioxide Photocatalysis. Journal of Photochemistry and Photobiology C Photochemistry Reviews. 2000;**1**:1

[7] Tryk DA, Fujishima A, Honda K. Recent Topics in Photoelectrochemistry: Achievements and Future Prospects. Electrochimica Acta. 2000;**45**:2363

[8] Grätzel M. Photoelectrochemical Cells. Nature. 2001;**414**:338

[9] Hagfeldt A, Grätzel M. Light-induced Redox Reactions in Nanocrystalline Systems. Chemical Reviews. 1995;**95**:49

[10] Linsebigler AL, Lu G, Yates JT Jr. Photocatalysis on TiO_2 Surfaces: Principles Mechanisms and Selected Results. Chemical Reviews. 1995;**95**:735

[11] Millis A, Le Hunte S. An Overview of Semiconductor Photocatalysis. Journal of Photochemical and Photobiology. A. 1997;**108**:1

[12] Alivisatos AP. Perspectives on the Physical Chemistry of Semiconductor Nanocrystals. The Journal of Physical Chemistry. 1996;**100**:13226

[13] Alivisatos AP. Semiconductor Clusters, Nanocrystals, and Quantum Dots. Science. 1996;**271**:933

[14] Burda C, Chen X, Narayanan R, El-Sayed MA. Chemistry and Properties of Nanocrystals of Different Shapes. Chemical Reviews. 2005;**105**:1025

[15] Murray CB, Kagan CR, Bawendi MG. Synthesis and Characterization of Monodisperse Nanocrystals and Close-Packed Nanocrystals Assemblies. Annual Review of Materials Science. 2000;**30**:545

[16] Roduner E. Nanoscopic Materials: Size-dependent Phenomena. Cambridge: Royal Society of Chemistry; 2006

[17] Kormann C, Bahnemann DW, Hoffmann MR. Preparation and Characterization of Quantum-Size Titanium Dioxide. The Journal of Physical Chemistry. 1988;**92**:5196

[18] Anpo M, Shima T, Kodama S, Kubokawa Y. Photocatalytic Hydrogenation of Propyne with Water on Small-Particle Titania: Size Quantization Effects and Reaction Intermediates. The Journal of Physical Chemistry. 1987;**91**:4305

[19] Sakai N, Ebina Y, Takada K, Sasaki T. Electronic Band Structure of

Titania Semiconductor Nanosheets Revealed by Electrochemical and Photoelectrochemical Studies. Journal of the American Chemical Society. 2004;**126**:5851

[20] Brinker CJ, Scherer GW. Sol–Gel Science. NY: Academic Press; 1990

[21] Hench LL, West JK. Sol–Gel Process. Chemical Reviews. 1990;**90**:33

[22] Aegerter MA, Mehrotra RC, Oehme I, Reisfeld R, Sakka S, Wolfbeis O, et al. Optical and Electronic Phenomena in Sol—Gel glasses and Modern Applications. Vol. 85. Berlin: Springer; 1996

[23] Hench LL, Wang SH, Nogues JL. In: Gunshor RL, editor. Multifunctional Materials. Vol. 878. Bellingham, WA: SPIE; 1988. p. 76

[24] Hench LL, Wilson MJR, Balaban C, Nogues JL. Sol–Gel Processing of Large Silica Optics. In: Proceedings of 4th International Conference on Ultrastructure Processing of Ceramics, Glasses, and Composites. Tucson, AZ; 1989

[25] Klein LC, Garvey GJ. In: Hench LL, Ulrich DR, editors. In: Ultrastructure Processing of Ceramics, Glasses and Composites. New York: Wiley; 1984. p. 88

[26] Szeifert JM, Fattakhova-Rohlfing D, Georgiadou D, Kalousek V, Rathousky J, Kuang D, et al. "Brick and Mortar" Strategy for the Formation of Highly Crystalline Mesoporous Titania Films from Nanocrystalline Building Blocks. Chemistry of Materials. 2009;**21**:1260-1265

[27] Gratzel M, Bach U, Lupo D, Comte P, Moser JE, Weissortel F, et al. Solid-State Dye- Sensitized Mesoporous TiO_2 Solar Cells with High Photon-To-Electron Conversion efficiencies. Nature. 1998;**395**:583-585

[28] Birkefeld LD, Azad AM, Akbar SA. Carbon Monoxide and Hydrogen Detection by Anatase Modification of Titanium Dioxide. Journal of the American Ceramic Society. 1992;**75**:2964-2968

[29] Terzian R. Photocatalyzed Mineralization of Cresols in Aqueous Media with Irradiated Titania. Journal of Catalysis. 1991;**128**:352-365

[30] Wang C, Yin L, Zhang L, Qi Y, Lun N, Liu N. Large Scale Synthesis and Gas-Sensing Properties of Anatase TiO_2 Three-Dimensional Hierarchical Nanostructures. Langmuir. 2010;**26**:12841-12848

[31] Hohn N, Schlosser SJ, Bießmann L, Song L, Grott S, Xia S, et al. Impact of Catalytic Additive on Spray Deposited and Nanoporous Titania Thin Films Observed via *in Situ* X-Ray Scattering: Implications for Enhanced Photovoltaics. ACS Applied Nanomaterials. 2018;**1**:4227-4235

[32] Green MA, Hishikawa Y, Warta W, Dunlop ED, Levi DH, Hohl-Ebinger J, et al. Solar Cell Efficiency Tables (version 50). Progress in Photovoltaics. 2017;**25**:668-676

[33] Ares AE. Editor. IntechOpen: Thin Films; 2021. DOI: 10.5772/intechopen.87838

[34] Roth SV, Santoro G, Risch JFH, Yu S, Schwartzkopf M, Boese T, et al. Patterned Diblock Co-Polymer Thin Films as Templates for Advanced Anisotropic Metal Nanostructures. ACS Applied Materials & Interfaces. 2015;**7**:12470-12477

[35] Peinemann K-V, Abetz V, Simon PFW. Asymmetric Superstructure Formed in a Block Copolymer Via Phase Separation. Nature Materials. 2007;**6**:992-996

[36] Tauster SJ, Fung SC, Garten RL. Strong Metal-Support Interactions. Group

8 Noble Metals Supported on Titanium Dioxide. Journal of the American Chemical Society. 1978;**100**:170-175

[37] Shastri AG, Datye AK, Schwank J. Gold-Titania Interactions: Temperature Dependence of Surface Area and Crystallinity of TiO_2 and Gold Dispersion. Journal of Catalysis. 1984;**87**:265

[38] Ragai J, Sing KSW, Mikhail R. Origin of Porosity in Titania Gels. I. Microporous and Mesoporous Gels prepared from Titanous Chloride and Ammonia. Journal of Chemical Technology and Biotechnology. 1980;**30**:1

[39] Campbell LK, Na BK, Ko EI. Synthesis and Characterization of Titania Aerogels. Chemistry of Materials. 1992;**4**(6):1329-1333

[40] Zhu Z, Tsung LY, Tomkiewicz M. Morphology of TiO_2 Aerogels. 1. Electron Microscopy. The Journal of Physical Chemistry. 1995;**99**:15945

[41] Dagan G, Tomkiewicz M. Preparation and Characterization of TiO_2 Aerogels for Use as Photocatalysts. Journal of Non-Crystalline Solids. 1994;**175**:294

[42] Ji L, Rong J, Yang Z. Opal Gel Templated Synthesis of Oblate Titania Opals. Chemical Communications. 2003;**1080**

[43] Avnir D, Levy D, Reisfeld R. The Nature of the Silica Cage as Reflected by Spectral Changes and Enhanced Photostability of Trapped Rhodamine 6G. The Journal of Physical Chemistry. 1984;**88**:5956-5959

[44] Nosaka Y, Tohriiwa N, Kobayashi T, Fujii N. Two Dimensionally Poled Sol–Gel Processing of Titania Film doped with Organic Compounds for Nonlinear Optical Activity. Chemistry of Materials. 1993;**5**:930-932

[45] Choi W, Termin A, Hoffmann MR. The Role of Metal Ion Dopants in Quantum-sized TiO_2: Correlation between Photoreactivity and Charge Carrier Recombination Dynamics. The Journal of Physical Chemistry. 1994;**98**:13669

[46] Li FB, Li XZ, Hou MF. Photocatalytic Degradation of 2-Mercaptobenzothiazole in Aqueous La3+--TiO_2 Suspension for Odor Control. Applied Catalysis B: Environmental. 2004;**48**:185-194

[47] Li W, Wang Y, Lin H, Shah SI, Huang CP, Doren DJ, et al. Bandgap Tailoring of Nd^{3+}-Doped TiO_2 Nanoparticles. Applied Physics Letters. 2003;**83**:4143

[48] Chen X, Lou Y, Dayal S, Qiu X, Krolicki R, Burda C, et al. Doped Semiconductor Nanomaterials. Nanosci. Nanotechnol. 2005;**5**:1408

[49] Umebayashi T, Yamaki T, Itoh H, Asai K. Analysis of Electronic Structures of 3d Transition Metal-Doped TiO_2 based on Band Calculations. Journal of Physics and Chemistry of Solids. 2002;**63**:1909

[50] Haidrya AA, Puskelova J, Plecenik T, Durina P, Gregus J, Truchly M, et al. Characterization and Hydrogen Gas Sensing Properties of TiO_2 Thin Films Prepared by Sol–Gel Method. Applied Surface Science. 2012;**259**:270

[51] Zheng C, Lin J, Song X, Gan Q, Lin X. TiO_2-Nanoparticle-Shelled Light-Driven Microcleaner for Fast and Highly Efficient Degradation of Organic Pollutants. ACS Applied Nano Materials. 2022;**5**:16573

[52] Lin J, Tao Y, Liu J, Zheng C, Song X, Dai P, et al. TiO_2 @ Carbon Microsphere Core-Shell Micromotors for Photocatalytic Water Remediation. Optical Materials. 2022;**124**:111989

[53] Dong R, Zhang Q, Gao W, Pei A, Ren B. Nanomotor lithography. ACS Nano. 2016;**10**:839

[54] Moo JG, Pumera M. Chemical Energy Powered Nano/micro/macromotors and the Environment. Chemistry--A European Journal. 2015;**21**:58-72

[55] Bertucci S, Megahd H, Dodero A, Fiorito S, Di Stasio F, Patrini M, et al. Mild Sol–Gel Conditions and High Dielectric Contrast: A Facile Processing toward Large-scale Hybrid Photonic Crystals for Sensing and Photocatalysis. ACS Applied Materials & Interfaces. 2022;**14**:19806-19817

[56] Chen Y, Crittenden JC, Hackney S, Sutter L, Hand DW. Preparation of a Novel TiO_2-based p-n Junction Nanotube Photocatalyst. Environmental Science & Technology. 2005;**39**:1201

[57] Lee S, Jeon C, Park Y. Fabrication of TiO_2 Tubules by Template Synthesis and Hydrolysis with Water Vapor. Chemistry of Materials. 2004;**16**:4292

[58] Qiu JJ, Yu WD, Gao XD, Li XM. Sol–Gel Assisted ZnO Nanorod Array Template to Synthesize TiO_2 Nanotube Arrays. Nanotechnology. 2006;**17**:4695

[59] Limmer SJ, Seraji S, Forbess MJ, Wu Y, Chou TP, Nguyen C, et al. Electrophoretic Growth of Lead Zirconate Titanate Nanorods. Advanced Materials. 2001;**13**:1269

[60] Limmer SJ, Seraji S, Forbess MJ, Wu Y, Chou TP, Nguyen C, et al. Template-Based Growth of Various Oxide Nanorods by Sol–Gel Electrophoresis. Advanced Functional Materials. 2002;**12**:59

[61] Cao GZ. Growth of Oxide Nanorod Arrays through Sol Electrophoretic Deposition. The Journal of Physical Chemistry. B. 2004;**108**:19921

[62] Miao Z, Xu D, Ouyang J, Guo G, Zhao Z, Tang Y. Electrochemically Induced Sol–Gel Preparation of Single-Crystalline TiO_2 Nanowires. Nano Letters. 2002;**2**:717

[63] Natarajan C, Nogami G. Cathodic Electrodeposition of Nanocrystalline Titanium Dioxide Thin Films. Journal of the Electrochemical Society. 1996;**143**(5):1547

[64] Sugimoto T, Okada K, Itoh H. Synthesis of Uniform Spindle-Type Titania Particles by the Gel-Sol Method. Journal of Colloid and Interface Science. 1997;**193**:140

[65] Sugimoto T, Zhou X. Synthesis of Uniform Anatase TiO_2 Nanoparticles by The Gel-Sol Method 2. Adsorption of OH^- Ions to $Ti(OH)_4$ Gel and TiO_2 Particles. Journal of Colloid and Interface Science. 2002;**252**:347

[66] Sugimoto T, Zhou X, Muramatsu A. Synthesis of Uniform Anatase TiO_2 Nanoparticles by The Gel-Sol Method 1. Solution Chemistry of Ti(OH)(4-n)+ (n) Complexes. Journal of Colloid Interface Science. 2002;**252**:339

[67] Sugimoto T, Zhou X, Muramatsu A. Synthesis of Uniform Anatase TiO_2 Nanoparticles by The Gel-Sol Method. 4. Shape Control. Journal of Colloid and Interface Science. 2003;**259**:53

[68] Sugimoto T, Zhou X, Muramatsu A. Synthesis of Uniform Anatase TiO_2 Nanoparticles by The Gel-Sol Method. 3. Formation Process and Size Control. Journal of Colloid and Interface Science. 2003;**259**:43

[69] Uekawa N, Kajiwara J, Kakegawa K, Sasaki Y. Low Temperature Synthesis and Characterization of Porous Anatase TiO_2 Nanoparticles. Journal of Colloid and Interface Science. 2002;**250**:285

[70] Fujii H, Inata K, Ohtaki M, Eguchi K, Arai H. Synthesis of CdS/TiO_2 Nanocomposite via TiO_2 coating on CdS Nanoparticle by Compartmentalized Hydrolysis of Ti alkoxide. Journal of Materials Science. 2001;**36**:527

[71] Matsumoto H, Matsunaga T, Sakata T, Mori H, Yoneyama H. Size Dependent Fluorescence Quenching of CdS Nanocrystals Caused by TiO_2 Colloids as a Potential-Variable Quencher. Langmuir. 1995;**11**:4283

[72] Qian X, Qin D, Bai Y, Li T, Tang X, Wang E, et al. Photosensitization of TiO_2 Nanoparticulate Thin Film Electrodes by CdS Nanoparticles. Journal of Solid State Electrochemistry. 2001;**5**:562

[73] Shen Q, Arae D, Toyoda T. Photosensitization of Nanostructured TiO_2 with CdSe Quantum Dots: Effects of Microstructure and Electron Transport in TiO_2 Substrates. Journal of Photochemical and Photobiology. A. 2004;**164**:75

[74] Spanhel L, Weller H, Henglein A. Photochemistry of Semiconductor Colloids. 22. Electron Ejection from Illuminated CdS into attached Titanium and Zinc Oxide particles. Journal of the American Chemical Society. 1987;**109**:6632

[75] Hoyer P, Koenenkamp R. Photoconduction in Porous TiO_2 Sensitized by PbS Quantum Dots. Applied Physics Letters. 1995;**66**:349

[76] Fitzmaurice D, Frei H, Rabani J. Time-Resolved Optical Study on the Charge carrier Dynamics in a TiO_2/AgI Sandwich Colloid. The Journal of Physical Chemistry. 1995;**99**:9176

[77] Desilvestro J, Gratzel M, Kavan L, Moser J, Augustynski J. Highly Efficient Sensitization of Titanium Dioxide. Journal of the American Chemical Society. 1985;**107**:2988

[78] Vogel R, Hoyer P, Weller H. Quantum-sized PbS, CdS, Ag2S, Sb2S3, and Bi2S3 Particles as Sensitizers for Various Nanoporous wide-Bandgap Semiconductors. The Journal of Physical Chemistry. 1994;**98**:3183

[79] Robel I, Subramanian V, Kuno M, Kamat PV. Quantum Dot Solar Cells. Harvesting Light Energy with CdSe Nanocrystals Molecularly Linked to Mesoscopic TiO_2 Films. Journal of the American Chemical Society. 2006;**128**:2385

[80] Tian Y, Tatsuma T. Mechanisms and Applications of Plasmon-induced Charge Separation at TiO_2 Films Loaded with Gold Nanoparticles. Journal of the American Chemical Society. 2005;**127**:7632

[81] Tian Y, Tatsuma T. Plasmon-induced Photoelectrochemistry at Metal Nanoparticles Supported on Nanoporous TiO_2. Chemical Communications. 2004;**1810**

[82] Cozzoli PD, Curri ML, Agostiano A. Efficient Charge Storage in Photoexcited TiO_2 Nanorod-Noble Metal Nanoparticle Composite Systems. Chemical Communications. 2005;**3186**

[83] Bottein T, Bouabdellaoui M, Jean-Benoît C, Favre L, David T, Putero M, et al. Large Scale Self-Organization of 2D Hexagonal Ge and Au Nanodots on patterned TiO_2. ACS Applied Nano Materials. 2019;**2**:2026-2035

[84] Lim SH, Luo J, Zhong Z, Ji W, Lin J. Room Temperature Hydrogen Uptake by TiO_2 Nanotubes. Inorganic Chemistry. 2005;**44**:4124

[85] Bavykin DV, Lapkin AA, Plucinski PK, Friedrich JM, Walsh FC. Reversible Storage of Molecular Hydrogen

by Sorption into Multilayered TiO_2 nanotubes. The Journal of Physical Chemistry. B. 2005;**109**:19422

[86] Adachi M, Murata Y, Okada L, Yoshikawa S. Formation of Titania Nanotubes and Applications for Dye-sensitized Solar Cells. Journal of the Electrochemical Society. 2003;**150**:G488

[87] Ohsaki Y, Masaki N, Kitamara T, Wada Y, Okamoto T, Sekino T, et al. Dye-Sensitized TiO_2 Nanotube Solar Cells: Fabrication and Electronic Characterization. Physical Chemistry Chemical Physics. 2005;**7**:4157

[88] Mor GK, Shankar K, Paulose M, Varghese OK, Grimes CA. Use of Highly-Ordered TiO_2 Nanotube Arrays in Dye-Sensitized Solar Cells. NanoLett. 2006;**6**:215

[89] Luo H, Takata T, Lee Y, Zhao J, Domen K, Yan Y. Photocatalytic Activity Enhancing for Titanium Dioxide by Co-Doping with Bromine and Chlorine. Chemistry of Materials. 2004;**16**:846

[90] Jing D, Zhang Y, Guo L. Study on the synthesis of Ni-Doped Mesoporous TiO_2 and its Photocatalytic Activity for Hydrogen Evolution in Aqueous Methanol Solution. Chemical Physics Letters. 2005;**415**:74

[91] Yang SG, Quan X, Li XY, Fang N, Zhang N, Zhao HM. J. Environ. Sci. Vol. 17. (Beijing, China); 2005. p. 290

[92] Bonhote P, Gogniat E, Campus F, Walder L, Grätzel M, Nanocrystalline Electrochromic Displays. Displays 1999; 20: 137

[93] Bonhote P, Gogniat E, Gra¨tzel M, Ashrit PV. Novel Electrochromic Devices Based on Complimentary Nanocrystalline TiO_2 and WO3 Thin Films. Thin Solid Films. 1999;**350**:269

[94] Singh JP, Singh V, Sharma A, Pandey G, Chae KH, Lee S. Approaches to Synthesize MgO Nanostructures for Diverse Applications. Heliyon. 2020;**6**:e04882

Chapter 4

Synthesis of Some Functional Oxides and Their Composites Using Sol-Gel Method

Arafa Hassen, Adel M. El Sayed, Azza Al-Ghamdi and Mohamed Shaban

Abstract

Two main approaches for nanomaterials fabrication are the top-down and the bottom-up methods. The first is limited to mechanical grinding, thermal evaporation, ion sputtering, arc discharge, pulsed laser ablation, and other physical and chemical vapor deposition. These routes are costly, consume higher energy, and require complex technology such as ultrahigh vacuum. The bottom-up methods refer to the production of complex nanostructured materials from atoms and molecules. This approach is relatively simple and low in cost. However, it requires a good knowledge of the optical properties of the particles and their modifications when the particles are integrated with nanostructures. One of the widest bottom-up methods is the sol-gel. It involves a solution or sol (single-phase liquid) that undergoes a sol-gel transition (stable suspension of colloidal particles). In this chapter, we throw light on the history of sol-gel, its advantages, and limitations, operating this method for the production of different types of nanomaterials in the form of powders or thin films. In addition, some applications of the sol-gel-derived nanosized materials will be discussed.

Keywords: sol-gel preparation, metal oxide nanomaterials, characterization, oxides, sol-gel chemistry

1. Introduction

Nanosized material, a material with at least one dimension limited to ˂100 nm (A nanometer is 10^{-9} of a meter.), displays unique and unexpected physicochemical properties. This behavior of nanomaterials arises from the large surface area to volume ratio and the quantum confinement effect that can be defined as the reduction of the band structure of the material into discrete quantum levels and the emerging of new energies for the electrons, resulting from the limited size of its particle, also known as the "size-effect." **Figure 1** shows that the surface atoms/volume ratio increases exponentially with decreasing particle size. Increasing the surface of the material increases its reactivity and photoelectrochemical performance. The accumulation of information on nanosized materials resulted in or emerged two branches, "Nanoscience" and "Nanotechnology." The former focuses on the preparation and

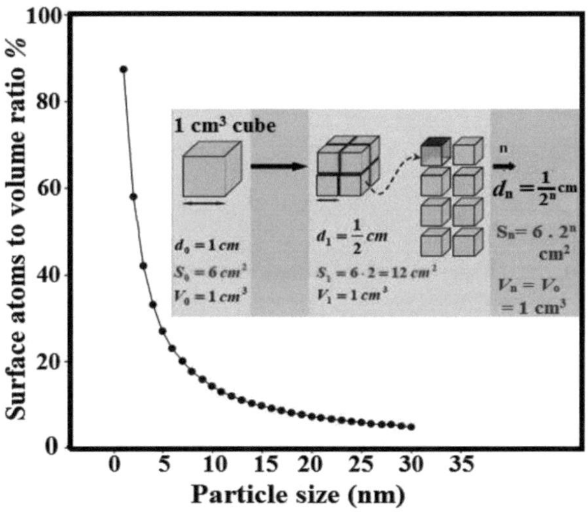

Figure 1.
The surface atoms/volume ratio (the determined surface-to-bulk atomic ratio) [1].

characterization of the nanomaterials and the fundamental study of their properties, whereas the latter is related to designing and using structures and devices based on these nanosized materials in different applications [2, 3].

The literature survey revealed that the physical and chemical properties of nanosized materials as well as the particles' morphology (0D, 1D, 2D, ...), also depend on the preparation method and preparative parameters and conditions. With the continuous headway of nanotechnology, there are several methods or techniques for preparing nanosized materials which can be classified into two main branches; top-down and bottom-up; the top-down methods are based on breaking down large pieces/particles of the material to convert it to the required nanostructures. The "bottom-up" methods are based on assembling single atoms/molecules (in solutions or gas phase) into larger nanostructures. We will discuss the details of one of the bottom-up methods in this chapter, named the sol-gel.

2. Sol-gel chemistry

As a phenomenon, the sol-gel transition was discovered and explored by Ebelmen in 1846 by observing the slow transformation of silicic esters, in the presence of moisture, to hydrated silica and the spontaneous gelation when the alkoxide was placed in contact with the atmosphere [4]. However, the interest in the sol-gel method began in 1980 and received a continuous and increased interest exponentially until today, and we expect a growing interest during the current decade, as shown in **Figure 2**.

A sol is defined as a colloidal system in which the dispersion medium is a liquid, and the dispersed phase is a polymerized molecule or fine particles, where the particle/molecular size should be in the range of 1 nm – 1 μm. A gel is a continuous solid network that supports the continuous liquid phase [5]. In the typical sol-gel process, consecutive steps are the sol formation through hydrolysis, the sol-gel transition (gel state), the gel drying, and conversion into a calcined material, as shown in **Figure 3**. The chelating agent binds tightly with the metal ions to prevent the formation of

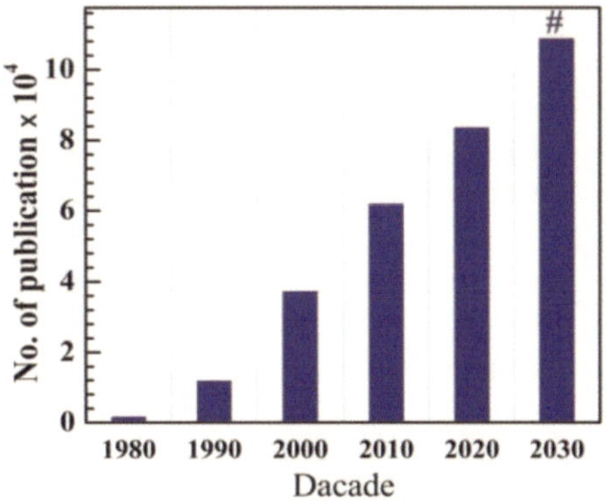

Figure 2.
Number of publications/decade utilizing the sol-gel route (Scopus database), # is the expected number [Scopus database].

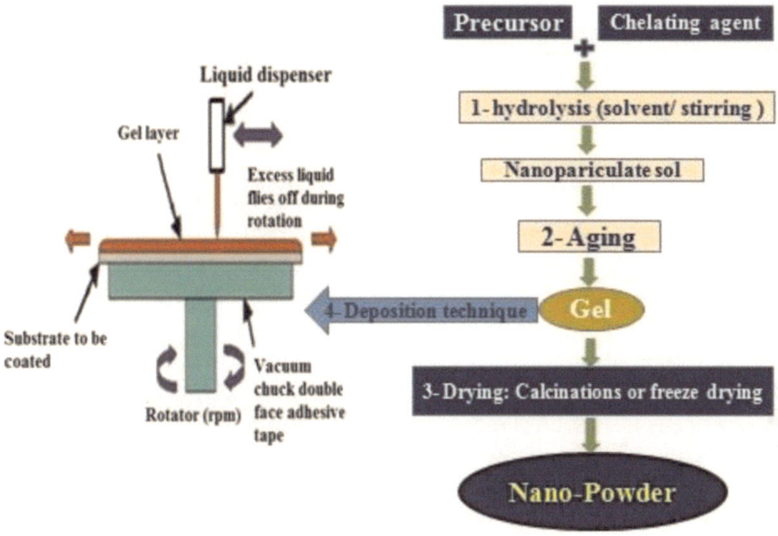

Figure 3.
Steps of sol-gel chemistry for powder and thin film formation [6, 7].

aggregations. The sol-gel chemistry begins with mixing the precursor (acetate, nitrate, or chloride) with the solvent. If water is the solvent, the sol-gel is hydrolytic but named nonhydrolytic sol-gel in the case of using an organic solvent such as ethanol [6]. The solution prepared by the sol-gel chemistry is used cooperatively with coating techniques such as spray, dip, and spin coating. For thin film deposition, the chelating agent has the role of stabilizer to prevent the metal ions to be precipitated or agglomerated. The spin-coated films will form in nanoparticulate layers, as will be discussed.

According to Brinker and others, the sol-gel method is a technology where the solution containing the precursor solid materials evolves gradually to form a networked gel comprising both the liquid and solid phases. The precursors react with each other in the common solvent to form a colloidal suspension (sol). This sol undergoes a hydrolysis reaction that could be represented as $M(OR)_n + nH_2O \rightarrow M(OH)_n + nROH$ and condensation reaction: $M(OH)_n \rightarrow MO_{0.5n} + (0.5n)H_2O$ to form a continuous liquid network (gel) [4, 7]. Once the gel is formed, it can be coated and dried to form thin films, as will be discussed, or be further dried, and densified at higher temperatures to fine powder, depending on the application [5], see **Figure 3**.

The sol-gel approach became one of the key technologies of the twenty-first century owing to low-energy consumption, reproducibility, eco-friendly, simplicity, low-cost, and pollution-free. In addition, it allows the combination of inorganic/organic materials in a single-phase and yields an organic/inorganic hybrid coating which attracted great attention owing to their high compatibility, good adhesion to the substrate, and corrosion resistance. Moreover, the sol-gel technology is represented in low-temperature requirements, repeatability, and controllability. In addition, it is possible to tune the intrinsic properties and the elemental chemical composition of the material. The final product of the sol-gel reaction can be controlled by precursors, pH, processing time, and molar ratios between the reacting agents. Löbmann revealed that the sol-gel route could yield various topologies; porous $\lambda/4$ films, dense interference layers, and arrays of antireflective structures (called moth-eye). These topologies can be used for antireflective coatings for architectural glazing, the display industry, solar energy conversion, and ophthalmic lenses [8]. Controlling the structure of sol-gel prepared film could yield highly selective gas sensors [9]. Chen et al. [10] studied the effect of pH value (1–10) on the corrosion protection ability of the sol-gel coatings. The highest condensation degree occurred at pH 4, resulting in a compact and stable 3D sol-gel network of high crosslinking density, and this provided highly effective corrosion protection.

It was also found that the photocatalytic properties of the sol-gel prepared TiO_2 nanopowder depend mainly on the sol composition, where the addition of water, HCl, and diethanolamine as well as the type of alcohol as solvent (ethanol, propanol, and butanol) were found to greatly affect the photocatalytic activity of the powder toward bromophenol blue dye removal [11]. Luo et al. [12] studied some of the variables related to the sol-gel preparation of CaO as a high-performance sorbent and they concluded that the molar ratio of $H_2O:Ca^{2+}$ had a minor effect on the CO_2 sorption performance of the CaO, and the optimal molar ratio was 80:1. The optimal molar ratio of citric acid: calcium nitrate optimal molar ratio is 1:1, and adding an excess of citric acid led to more gaseous products. In addition, when the pH 3, the sol-gel structure was destroyed, and the optimal pH value was 2, where the best performance of CaO sorbent was achieved. A. C.-Soria et al. [13] fabricated Fe_3C/few-layered graphene core/shell nanoparticles, with potential magnetic properties, embedded in a carbon matrix by a modified two-step surfactant sol-gel method, where the hydrolysis, polycondensation, and drying took place in a one-pot. Hashjin et al. [14] tuned up the sol-gel technique for preparing high-durable superhydrophobic coatings. The prepared layers are useful for anti-icing, self-cleaning, and anti-bacterial applications, in the energy and photovoltaic devices, textile and coating industry, construction, and aerospace industry.

Sol-gel technique, among various solution methods, is found to be more suitable for metal oxide thin films and nanopowder. Controlling the conditions of preparation, nanoparticles of controlled shape/morphology, control stoichiometry, size, textural, surface characteristics, purity, and high quality can be obtained. Besides, uncomplicated ideas can be executed via this technique for more recent and advanced technological applications [6]. In the following section, some examples of the sol-gel derived nanostructures will be mentioned with their characterization and some related applications. The data presented here are based on our experimental results. It would be better to throw light on some selected materials that were prepared using the sol-gel method.

3. Sol-gel preparation of NiO, CdO, SnO_2, and PbO and their nanocomposites

3.1 Experimental (preparation and characterization techniques)

The precursor materials used for NiO, CdO, SnO_2, and PbO preparation were: $NiCl_2{\cdot}6H_2O$ of molecular weight (M_W = 237.7), supplied by Schorlau, Spin, $Cd(NO_3)_2$ M_W = 236.42, supplied by Nova Oleochem Limited, $SnCl_2.2H_2O$, M_W = 225.63, from Merck, and $CH_3COO)_2Pb.3H_2O$, M_W = 279.33, Adwik, Egypt, were used to prepare 0.7 M solutions by dissolving the required mass of each salt in 100 ml double-distilled water. To each solution, 8.825 g of oxalic acid ($C_2H_2O_4$), as a chelating agent, was added under stirring at 60°C for 1 h. The obtained solutions were maintained in an oven at 80–90°C for 20 h to evaporate the excess water above the precipitate. The solutions were then cooled to room temperature and aged for 24 at room temperature (RT). Finally, the four gel was calcined at 400°C for 2 h to obtain the nanopowders: NiO, CdO, $SnO_{2,}$ and PbO nanoparticles (NP). The characterization of these nanometal oxides will be discussed.

The identifying of the crystalline phase and samples purity was done by recording XRD spectra using the PANalytical X'Pert PRO diffractometer, with Cu K_α radiation of wavelength λ = 1.541 Å, and scan in the range of 2θ = 5.0–80°. High-resolution transmission electron microscopy (HR-TEM) of model JEM, 2100, Jeol, Japan, was used to check the particle size and morphology of the prepared materials. For polymer nanocomposite films, the surface morphology was evaluated using field emission-scanning electron microscopy FE-SEM (Carl ZEISS Sigma 500 VP). In addition, the UV–vis spectra in the wavelength range of 200–1600 nm were recorded on a Shimadzu spectrophotometer (UV-3600 UV-Vis–NIR) with an accuracy of ±0.2 nm.

3.2 Results and discussion

The crystallite size and phase identification of NiO, CdO, SnO_2, and PbO were examined by XRD, shown in **Figure 4**, and the shape and particle morphology was studied by HR-TEM, as shown in **Figure 5**. **Figure 4** shows the XRD pattern of NiO; the sharp peaks indicate that a good crystallite material was grown by the sol-gel technique. The diffraction peaks at 2θ = 37.14, 43.13, and 62.89° are indexed for the crystal planes (111), (200), and (220) of NiO of rhombohedral [*fcc* (face-centered cubic) with a lattice constant *a* = 4.175 Å], in agreement with JCPDS No. 44–1159. Scherrer's formula ($C_s = 0.9\lambda_{Cu}/\beta_{1/2} \times cos\ \theta$) was utilized to calculate the crystallite size (C_s), where $\beta_{1/2}$ is the full width at half maximum intensity. Considering the main detected peaks, the average C_s was ≈ 28 nm.

In the XRD pattern of the sol-gel prepared CdO nanoparticles, all of the detected diffraction peaks are indexed to the cubic phase of CdO with a lattice parameter a = 4.69483 Å. The peaks at $2\theta \approx$ 33.07°, 38.39°, 55.38°, 66°, and 69.38° are assigned to the (110), (200), (220), (311), and (222) crystal plans, respectively, according to JCPDS file No. 75–0592. This confirms the formation of CdO with excellent crystallinity and high purity, as no secondary phases were observed in the pattern of the CdO nanoparticles. The C_s of the CdO nanoparticles were found to be in the range of 66.4–73.2 nm with an average of 70.18 nm. The pattern of SnO_2 is also shown in **Figure 4**, where all the diffraction peaks with Miller indices of (110), (101), (200), (211), (220), (310), (112), and (301) are of the tetragonal (rutile) crystalline phase of

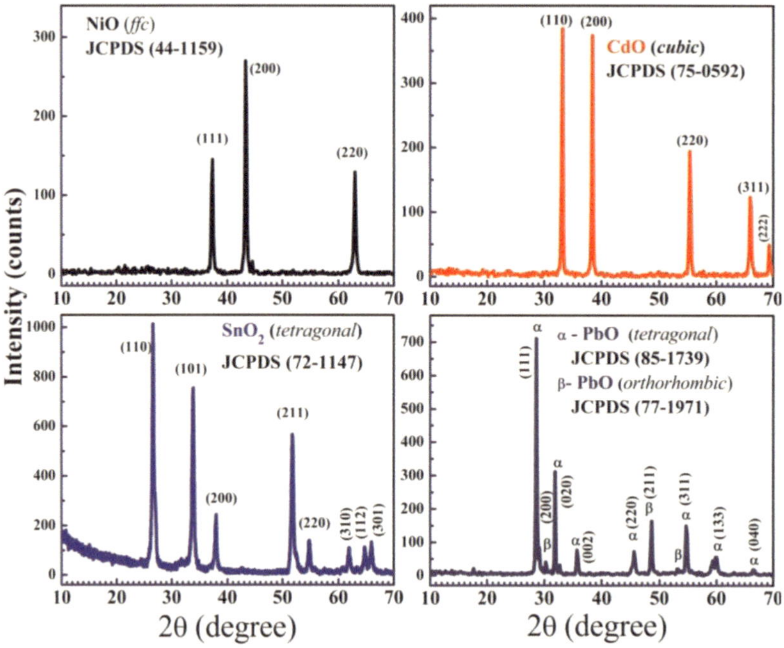

Figure 4.
(a-d) XRD patterns of the sol-gel-derived NiO, CdO, SnO_2, and PbO nanoparticles.

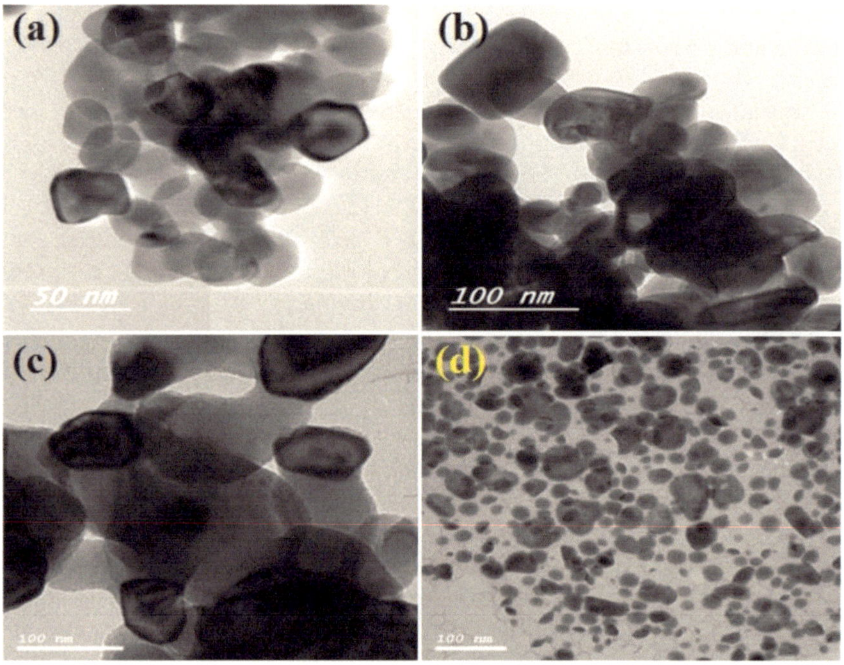

Figure 5.
HR-TEM of the sol-gel-derived NiO (a), CdO (b), SnO_2 (c), and PbO (d) nanoparticles.

SnO_2 according to JCPDS No. 72–1147. The (110) plane exhibits the highest intensity and presents the least surface energy and is the most thermodynamically and electrostatically stable [15]. The lattice parameters a, b, and c of the tetragonal SnO_2 phase are determined from the formula: $\frac{1}{d^2} = \frac{(h^2+k^2)}{a^2} + \frac{l^2}{c^2}$, and the calculated values were $a = b = 4.473$ Å and $c = 3.189$ Å. The C_s values are in the range of 17.93–47.06 nm with an average size of 30.2 nm. In the case of lead monoxide, the XRD pattern is a mixture of α- and β-PbO. The diffraction peaks of the orthorhombic β-PbO are at $2\theta = 29.06°$, 30.35°, 32.59°, and 53.17°, with $a = 5.88$ Å, $c = 4.74$ Å, according to JCPDS card No. 77–1971. The other peaks are assigned to the tetragonal α-PbO, with $a = 3.97$ Å, $c = 5.024$ Å, in agreement with JCPDS card No. 85–1739. The C_s of PbO is in the range of 24.4–113.4 nm with an average of 58.6 nm.

The HR-TEM image of NiO shows an average particle size of NiO in the range of 24.85–34.10 nm, which is smaller than that reported for NiO prepared from the thermal decomposition of $Ni(OH)_2$ at 600°C [16]. TEM image of the CdO shows that CdO nanoparticles are well-defined and their size is in the range of 52–116 nm with an average particle size of 72 nm. Besides, the image for the SnO_2 shows that SnO_2 grains are segregated together and form agglomerates or clusters of primary crystallites. Most of the observed particles are tetragonal in shape. The average particle size measured by HR-TEM is ~41 nm. Finally, the TEM image of the PbO formed as nanoparticles of sizes from tens of nm to <100 nm, with an average of about 59 nm, which is consistent with the XRD results.

Nickel oxide (NiO) is an interesting ceramic material with reasonable photostability and thermal stability, high melting at 1955°C, and a refractive index of ≈ 2.2. In addition, NiO is a p-type semiconductor with a wide optical bandgap (= 3.4–4 eV) [17, 18]. When the sol-gel prepared NiO was incorporated at 0.5 and 1.0 wt% into a polymer matrix composed of carboxymethyl cellulose–polyvinyl pyrrolidone (CMC–PVP) blend, the reflectivity and refractive index of the blend dramatically changed, as shown in **Figure 6**. For the pure blend, R% is in the range of 3–5 and decreases with increasing the incident wavelength. However, this behavior converted to a bell-like shape, and the R% increased to 6.5–13% after 1.0 wt% NiO doping. Similarly, the refractive index (n) value of the blend changed from 1.45 to 1.59 in the

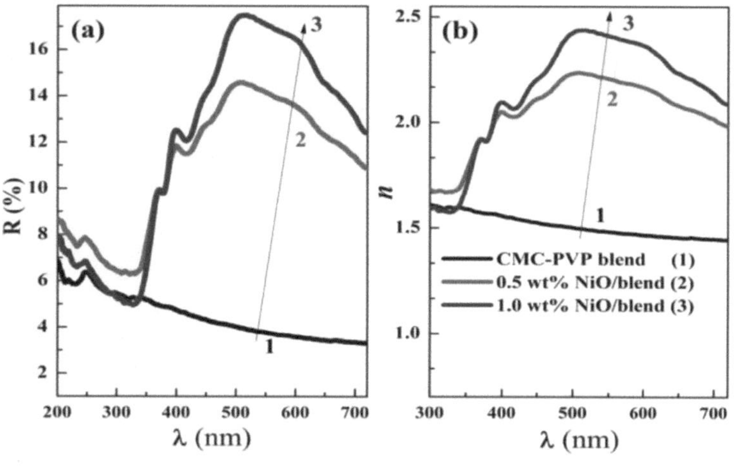

Figure 6.
Influence of NiO on the reflection R% (a), and refractive index n, (b) of a polymer blend.

visible region and increased significantly to 2.233 after the NiO loading. This illustrates that NiO increased the packing density of the blend, and these nanoparticles act as scattering centers to increase the dispersion of light and increase the reflection and reflectivity of the matrix [19]. Therefore, NiO/blend are suitable material for coatings, lenses, and for engineering, and optoelectronic applications [20].

Cadmium oxide (CdO) is a promising II–VI compound that has *n*-type semiconductivity, resistivity in the order of 10^{-2} –10^{-4} Ω cm [21], high optical transmittance in the visible region, and a refractive index of 2.49 [22]. The direct (indirect) band gap is in the range of 2.2–2.5 eV (1.36–1.98 eV) [23]. Therefore, CdO has been used for catalytic and sensing applications and in some optoelectronic devices [23]. The sol-gel prepared CdO when mixed with PVC polymer resulted in decreasing the transmittance of the polymer from 89% to the range of 70–84% and shrinking its band gap from 5.12 eV to 4.96 eV, as shown in **Figure 7**. This result may reflect the important applications of the CdO/PVC nanocomposites in optical and/or electrical devices [24].

Tin oxide (SnO_2) is also a transparent conducting oxide that exhibits outstanding electrical and optical properties. Its wide band gap (≈3.68 eV), high exciton binding energy (130 meV), high transmittance in the visible region of the spectra, high *n*-type conductivity (10^2–10^3 $\Omega^{-1}.cm^{-1}$) at ambient temperature, nontoxicity, thermal stability, chemical sensitivity, and the low-cost makes SnO_2 the best choice for the biomedical applications, gas sensing, photo-catalysis, solar cells, and future optoelectronic devices [25, 26]. In this chapter, we have used the sol-gel prepared SnO_2 nanoparticles as nanofillers to modify the optical properties of a ternary blend composed of carboxymethyl cellulose–polyethylene glycol–polyvinyl alcohol (CMC–PEG–PVA). As shown in **Figure 8**. The E_g value of the pure blend and SnO_2/CMC–PEG–PVA nanocomposite were calculated from the absorption spectra (*Abs.*) of the samples by using Tuac' relation: $(\alpha h\upsilon)^r = M(h\upsilon - E_g)$, where $h\nu$ is the energy of incident photons and r = 1/2 and 2 for the indirect and direct allowed transitions, respectively, and $\alpha = \frac{2.303\ Abs.}{\text{film thickness}}$ is the absorption coefficient.

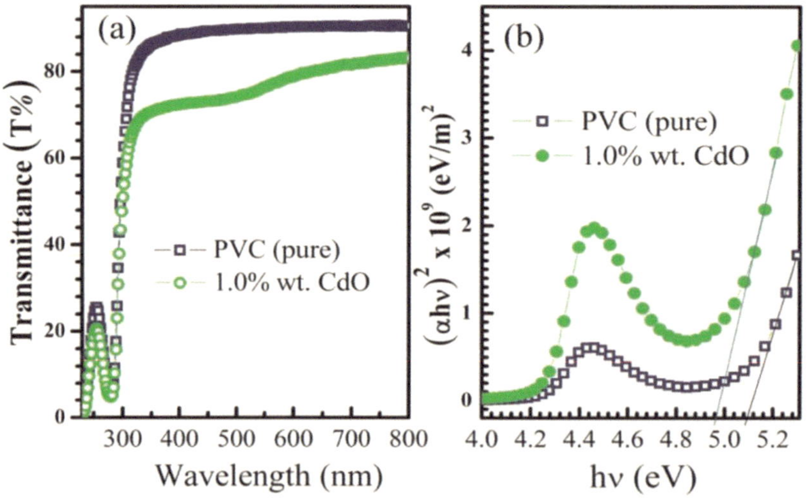

Figure 7.
Effect of 1.0 wt% CdO on the transmittance (a), and optical gap (b) of PVC polymer.

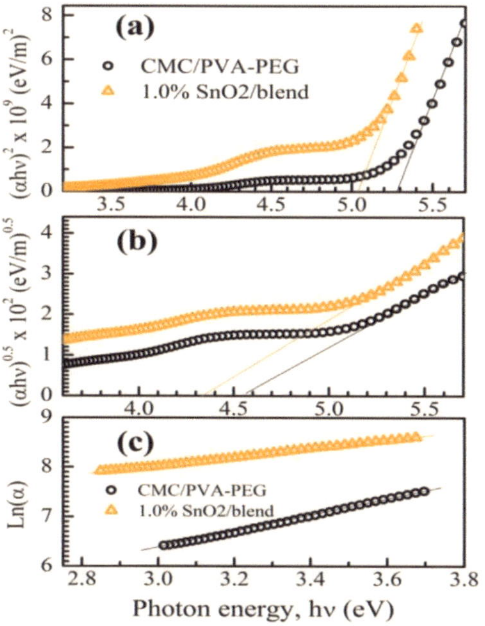

Figure 8.
Influence of SnO_2 nanoparticles (at 1.0 wt% loading) on the direct (a) and indirect (b) optical gap and Urbach energy (c) of CMC-PEG-PVA ternary blend.

As shown in **Figure 8**, both the direct and indirect transitions for the polymeric films are possible. This is evidenced by the linear relationship of both $(\alpha h\nu)^2$ and $(\alpha h\nu)^{0.5}$ on $h\nu$ at higher photon energies. Extra-plotting the straight-line portions of the curves to zero absorption gives the E_g values: direct E_g = 5.28 and 5.04 eV and the indirect E_g = 4.55 and 4.20 eV for the pure blend and 1.0 wt% SnO_2/blend nanocomposite, respectively. Introducing the SnO_2 nanoparticles induce energy levels inside the band gap of the blend matrix, resulting in the shrinking of E_g. Similarly, doping with α-Fe_2O_3 nanorods at 1.0 wt% decreased E_g of PVA–PEG from 5.28 to 4.83 eV [27]. The Urbach energy (E_U), which is the width of the exponential absorption edge, can be calculated using the following equation [28]: $\alpha = \alpha_o \exp\left[\frac{h\nu - E_c}{E_U}\right)$, where E_c and α_o are constants. The dependence of $ln\ (\alpha)$ on $h\upsilon$ for the films is shown in **Figure 8(c)**. The straight lines suggest that the absorption is according to the quadratic relation for inter-band transition, which satisfies the Urbach rule. The value of E_U is taken as the reciprocal of the slope of the lines (E_U = $(d(\ln\alpha)/dh\upsilon)^{-1}$), and its value was found to be 0.607 eV and 0.962 eV for pure and 1.0 wt% SnO_2 loaded film, respectively. Thus E_U changes inversely with E_g. Increasing the E_U is attributed to the disorder increase inside the material after SnO_2 nanoparticles incorporation, resulting in the tailing in the valence and conduction bands.

Lead oxides exist with a variety of oxidation states; α- and β-PbO, α- and β-PbO_2, Pb_2O_3, and Pb_3O_4 [29]. Among them, lead monoxide (PbO) is considered a transparent conducting oxide that has a high dielectric constant ε' = 25.9 [30] and a direct transition band gap of 1.96 eV for α-PbO [29, 31]. It is in use in a variety of applications, such as paints, pastes for a new lead acid battery, pigments, gas sensors, network modifiers in luminescent glassy materials, and nanodevices [31, 32]. Moreover, it can be used to increase the dielectric contestant and ac conductivity of polymeric

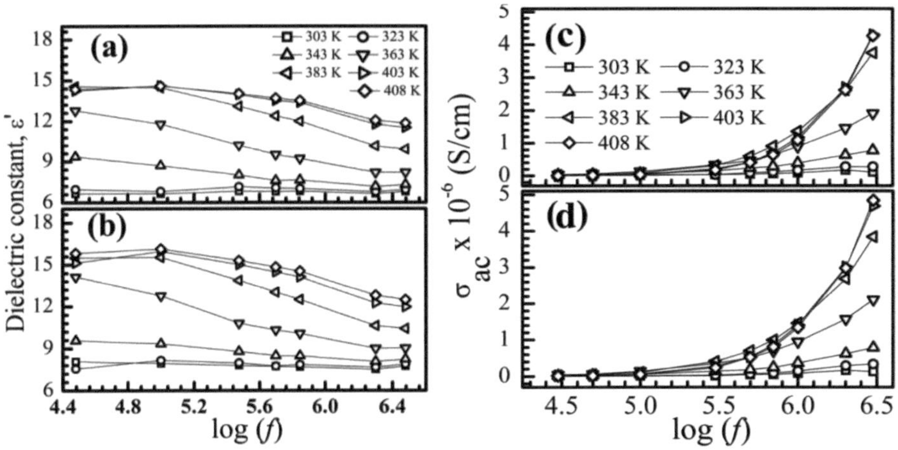

Figure 9.
Dielectric constant (a, b) and ac conductivity (c, d) of pure PVC and 1.0 wt % PbO-doped PVC.

materials such as PVC, as shown in **Figure 9**. Incorporation of PbO nanoparticles increases the interfacial polarization due to the heterogeneous structure inside PbO/PVC nanocomposites. Many conductive three-dimensional networks could be formed throughout the nanocomposite, assisting the charge carriers to hop from conducting clusters to neighbors and therefore increases the conductivity of the material [33].

4. Preparation of nanosized hematite with different sizes and morphology

Controlling the morphology of the material at the nanosize is the key to broadening its industrial applications. Here we will describe tuning the microstructure and morphology of the nanosized hematite (α-Fe_2O_3) by varying the oxalic acid (chelating agent) molar ratio from 0.0 to 2.0 and the annealing temperature in the range 350–750° C. α-Fe_2O_3 is a direct band gap (E_g = 2.0–2.2 eV), an n-type semiconductor that can absorb about 40% of the solar spectrum. It has several advantages, such as its abundance, high thermal stability, nontoxic, high resistance to corrosion, melting point of 1350°C, and a high specific capacitance of 3623 F/g. Therefore, α-Fe_2O_3 is widely used in several technological fields, including rechargeable Li-ion batteries, recording devices, catalysis, biomedical, optical devices, solar cells, and gas sensors [34–37]. The literature survey displays that The α-Fe_2O_3 in the form of 0 D and 1D nanostructures have grown increasing interest. For example, α-$Fe_2O_{3-\delta}$ nanoparticles showed enhanced electrochemical performance and cycling stability as anode materials for Li-ion batteries [38]. Moreover, hematite of plate-like morphology, displayed a significant hysteretic behavior at ambient temperature with saturation magnetization M_S = 2.15 emu/g and a coercivity H_C = 1140 Oe, remanent magnetization M_r = 0.125 emu/g [39].

4.1 The preparation and measurements

A nanopowder of α-Fe_2O_3 was prepared by a template-free sol-gel method as follows; 18.92 g of $FeCl_3.6H_2O$ (M_W = 270.3), supplied by Nova Oleochem Limited, was dissolved in 100 ml double distilled by the magnetic stirring for 1.0 h. Different amounts of oxalic acid were added to the solution to maintain the molar ratios of oxalic acid/$FeCl_3.6H_2O$ at 0.0:1.0, 0.5:1.0, 1.0:1.0, and 2.0:1.0. These samples named M = 0, M = 0.5, M = 1.0, and M = 2.0, respectively. The solution aged for about 15 h at the ambient temperature, then the excess water was thermally evaporated at 95°C for 3 h. The samples were calcined at 550°C in the air for 2.0 h. The sample M = 1.0 was calcined at different temperatures in the range of 350‾950°C, for 2 h. The crystal structural analysis and size and morphology of the prepared powders have been carried out by X-ray diffraction (using PANalytical's X'Pert PRO) and the HR-TEM (JEM 2100, Jeol).

4.2 Results and discussion

Figure 10 shows XRD patterns of the prepared materials (M = 0–2.0). All the diffraction peaks belong to the hexagonal structure of hematite according to JCPDS card no. 04–015-7029. No peaks related to any other FeO phases not detected. The strong peaks indicate the good crystallization of samples. The lattice parameters a and c and the volume $V = \frac{\sqrt{3}}{2}a^2c$ of the hexagonal cell are calculated from the following relation: $\frac{1}{d^2} = \frac{4}{3}\left(\frac{h^2+k^2+hk}{a^2}\right) + \frac{l^2}{c^2}$, where (hkl) is Miller's indices. Increasing M from 0.0 to 2.0 resulted in shifting the main peaks: (012), (104), and (110) to lower 2θ values, as shown in the inset of **Figure 10**.

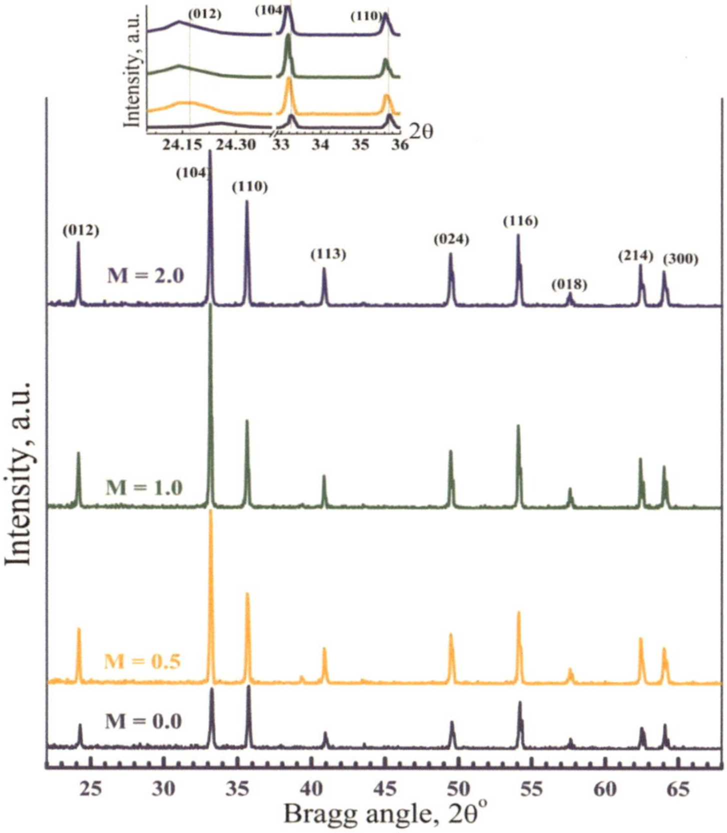

Figure 10.
XRD patterns of the sol-gel prepared hematite at different oxalic/hematite ratios.

This indicates increasing the *d*-spacing values of the material, where Bragg's law states that $d = n\lambda/2\,sin\theta$. The V value is increased from 299.62 Å^3 to 302.34 Å^3 with an increasing M value from 0.0 to 2.0. The average C_s were calculated considering the main reflections: (012), (104), (110), and (116) reflections. C_s was decreased from 95.7 to 61.1 nm with increasing M, which illustrates the role of oxalic acid as a chelating agent to slow down the nucleation rate and encourage the hinder the particles' agglomerations.

Figure 11 shows the XRD patterns of M = 1.0 sample calcined at 350–750°C. At 350°C, the thermal energy delivered to the material is sufficient to remove all of the organic molecules and the full oxidation of Fe into Fe_2O_3. Calcination temperatures higher than 350°C lead to an increase in the cell parameters and the diffraction peaks intensity. The C_s were increased from 51 to 73 nm., denoting the enhancement of the crystallinity with annealing. **Figure 12** shows the HR-TEM analysis of the sol-gel prepared hematite nanopowders at chelating agent/hematite molar ratios in the range of 0.0–2.0. As seen, the α-Fe_2O_3 nanocrystals aggregated more compactly at M = 0.0, a decrease of exposed surfaces of α-Fe_2O_3 nanocrystals is expected. Increasing M to 1.0 changes the morphology of the α-Fe_2O_3 structure to take the form of nanorods. However, increasing in M value to reach 2.0, these nanorods converted completely to nanoparticles. Therefore, the chelating agent ratio is a vital parameter in the nano synthesis process by sol-gel technique toward controlling the shape and morphology of α-Fe_2O_3 nanostructures. **Figure 13** displays the morphology of the sample M = 1.0 after calcination at different temperatures (350–750°C). The 350°C gives hematite nanopowder with nanorod/nanoneedle structure. The formation of the nanorods becomes clearer with increasing the calcination temperature to 750°C. The observed nanorods have average diameters of ~11.38 nm and different lengths. HR-TEM analysis illustrates that the shape/morphology of the obtained nanopowders is strongly influenced by the experimental conditions. Dissolving the $FeCl_3$. $6H_2O$ salt with oxalic acid at ambient temperature produces an iron oxalate ($FeC_2O_4.2H_2O$) solution. Increasing M value from 0.0 to 2.0 may change the pH value of the solution. The crystal structure of the intermediate phase ($FeC_2O_4.2H_2O$) [40], the pH value of the

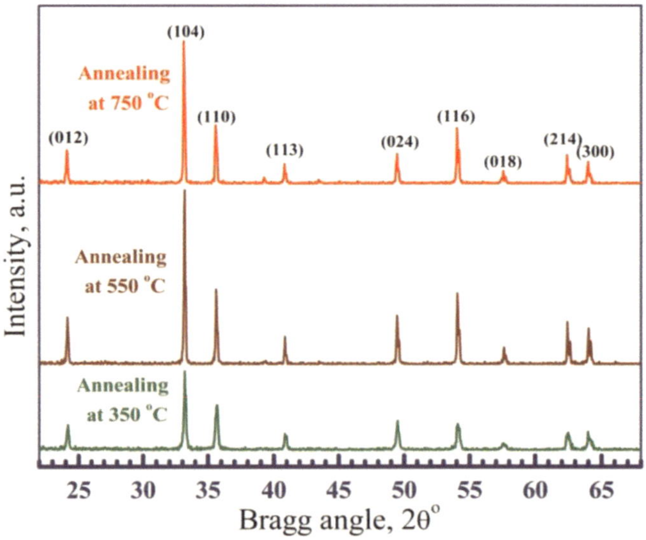

Figure 11.
XRD patterns of the sol-gel prepared hematite at different calcination temperatures.

Figure 12.
(a-d) HR-TEM images for the sol-gel prepared hematite at different oxalic/hematite ratios.

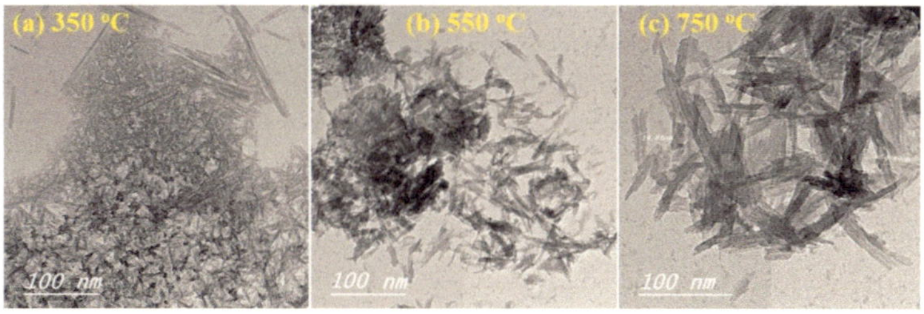

Figure 13.
(a-c) HR-TEM images for the sol-gel prepared hematite at different calcination temperatures.

solution, the aging time, and the calcination temperature and annealing rate may affect the shape/morphology of the obtained structure. Therefore, further investigations and studies on these parameters are required to understand the mechanism that leads to the growth of these different morphologies [41]. **Figure 14** depicts E-SEM of some prepared hexaferrites.

5. Ferrites and hexaferrites

The sol-gel method was also used to prepare different ceramics, for instance, as reported earlier [42, 43]. In which, M-type hexagonal ferrites $Ba_{1-x}Sr_xFe_{12}O_{19}$, where

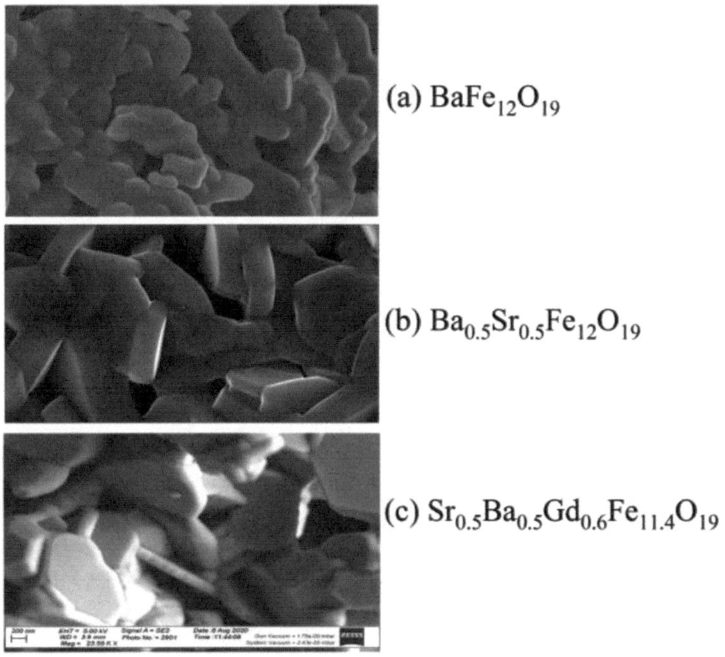

Figure 14.
(a-c) FE-SEM images of some prepared hexaferrites.

x ranged from 0 to 0.75, were synthesized [42]. The as-prepared ceramics were characterized by different techniques. The samples were single-phase based on the XRD and neutron diffractions. The average particle size of these hexaferrites was in the range of nanometers. Besides, the properties of the $Sr_{0.5}Ba_{0.5}RE_{0.6}Fe_{11.4}O_{19}$, where RE = La, Yb, Sm, Gd, Er, Eu, and Dy were reported [43]. It was emphasized that the quality of the samples was good based on the optimum use of the sol-gel method. **Figure 14** depicts the field emission-scanning electron microscope (FE-SEM) images of some prepared ceramic samples.

6. Rare earth oxides and titanium oxide-based perovskites

Lanthanum oxide (La_2O_3) is a rare earth sesquioxide that is optically active with a wide bandgap energy range of 4.3–5.4 eV. The ultrafine La_2O_3 NP has attractive properties for automobile exhaust-gas convectors, optical filters, and catalysis, as a strengthening agent in structural and ceramic materials, in high κ gate dielectric materials [44]. On the other hand, the Y_2O_3 is an interesting host material for solid-state lasers, high-temperature refractories (melting point of 2430°C), and infrared ceramics, owing to the distinctive *4f* electronic configuration, corrosion resistivity, and thermal stability [45]. On the other hand, TiO_2 is a semiconductor with $E_g \approx 3.21$ eV. Owing to its odorless, high melting (1843°C) and boiling (2972°C) points and high transmittance in the visible region, TiO_2 is used for various applications such as paint, sunscreen, and antireflective coating, food coloring, photocatalyst, photovoltaic in solar cells, and optical filter [46, 47]. TiO_2 has two crystalline forms; anatase and rutile. The first form is more chemically reactive and favorable in the industry [48]. Moreover, titanium oxide-based perovskites, such as $NaTiO_3$ (sodium titanate), attract increasing interest in PV cells and Li $^+$ batteries, as well as in biomedical and biochemical applications. This material has a band gap in the range of 2.2–2.5 eV and has a high cation-exchange capacity, thermal stability, and fast adsorption ability [49, 50].

6.1 Experimental

To prepare La_2O_3 nanoparticles (NP), 27.85 g of $LaCl_3.7H_2O$ (M_W = 371.37, Schorlau) were dissolved in 100 *ml* of distilled water (DW), and about 9.5 g of oxalic acid was added, and the solution was stirred at 55–60°C for 1.0 h. This sol was held at 90°C for 20 h, then cooled to room temperature (RT) and aged for one day at RT. After that, it was calcined at about 500°C for 2.0 h. To prepare Y_2O_3, about 9.6 g of Y $(NO_3)_3{\cdot}6H_2O$ (M_W = 383.01 g/mol, supplied by PubChem USA) was dissolved in 50 ml distilled water by stirring 20 min. Then 3.15 g of oxalic acid was added to this solution, and the stirring continued for 6 h. The solution was held at 90°C to get rid of the excess water. The sol was turned to gel by aging for ~12 h at RT, and then calcined at 400°C to obtain the nanopowders. For TiO_2 preparation, 24 mL of Ti [OCH $(CH_3)_2]_4$ (M_W = 284.22, from Sigma) was added to 60 mL of 2-propanol under stirring for 30 min, and several water drops were added for the hydrolysis process, and the stirring continued for 6 h. The sol was aged for ~12 h at RT, and then calcined at 400°C to obtain the nanopowders. The TiO_2 was used to prepare $NaTiO_3$ nanofibers, where 2.0 g of TiO_2 powder was mixed with 200 ml NaOH solution (10 M) by sonication for 1 h. The white solution was poured into a Teflon-lined autoclave at 45–50°C for 24 h. The $NaTiO_3$ precipitate was then washed several times with double-distilled water and then dried at 100°C for 48 h.

6.2 Results and discussion

Figure 15(a) displays the XRD pattern of the sol-gel-prepared La_2O_3. The sharp XRD peaks indicate the good crystallinity of the materials. The peaks at 2θ = 13.17, 26.48, 29.57, 40.11, 46.40, 51.55, 54.66, and 61.42° arise from the (001), (100), (011), (012), (110), (103), (112), and (202) crystal planes of La_2O_3 of hexagonal structure, consistent with the JCPDS no. 04–006-5083. The C_s was calculated considering the most intense peaks; the average value of C_s is ≈ 29.15 nm. The morphology was studied by the HR-TEM, as seen in the inset of the figure.

La_2O_3 has spherical NP morphology with a particle size of 30 nm, which is consistent with XRD results. Dal et al. [51] fabricated La_2O_3 NP of size 12.4 nm by sintering the La_2O_3 microparticles at 1250°C for 48 hrs with grinding for more than 3 hrs. This illustrates that our sol-gel process is low-cost in time and energy. **Figure 15(b)** displays the XRD peaks of the Y_2O_3, which are indexed as (211), (222), (400), (411), (332), (431), (440), (532), (622), (444), (552), and (800), corresponding to yttria (Y_2O_3) of body-centered cubic structured, according to JCPDS 043–1036. No other phases are detected in this spectrum, indicating that all the $Y(NO_3)_3$ entirely transformed into Y_2O_3 of single-phase after calcination at 400°C. The crystallite size of Y_2O_3 is 24.7 nm. The inset of **Figure 15(b)** shows the powder morphology, where the particle sizes of Y_2O_3 look like nanosheets (Ns), which are allocated with each other to be bigger than the calculated crystallite size from XRD. Similarly, Y_2O_3 Np of *Cs* = 8.7–27.8 nm) was prepared by the Pechini sol-gel method, and particle size measured by FE-SEM was in the range of 40–50 nm [52].

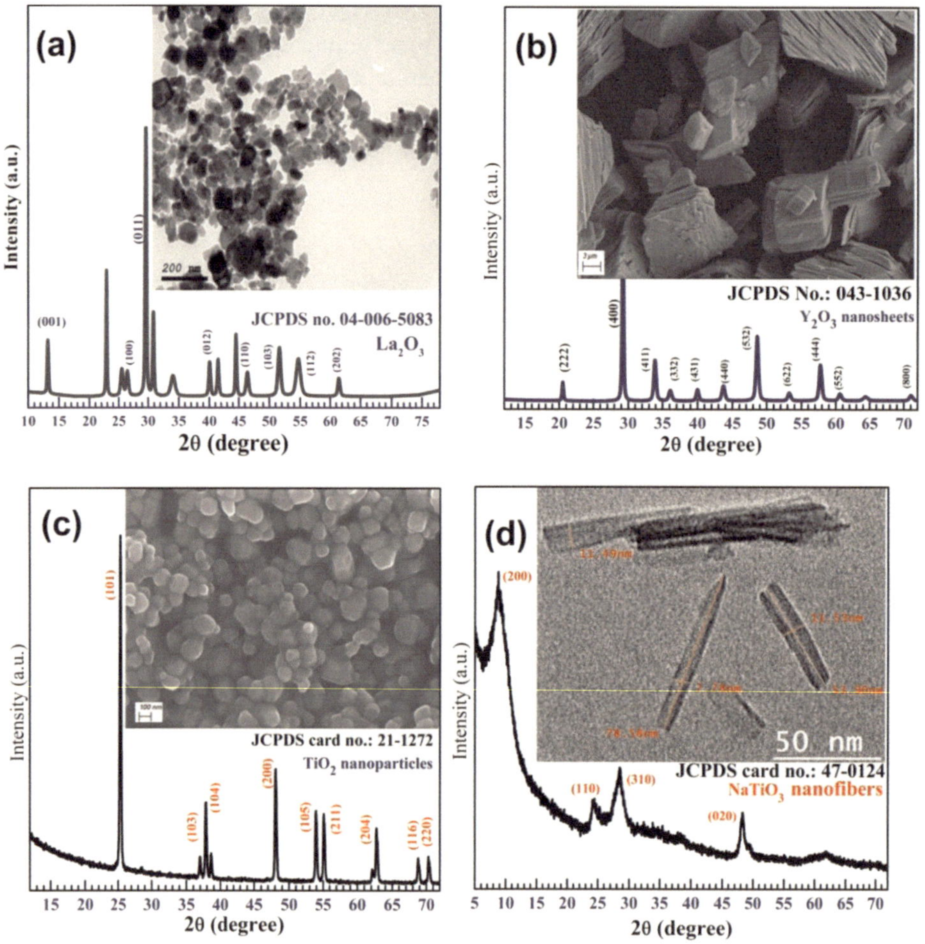

Figure 15.
(a-d) XRD patterns of La_2O_3, Y_2O_3, TiO_2, and $NaTiO_3$. The insets show the HR-TEM analysis of each material.

The XRD pattern of TiO_2, **Figure 15(c)**, consists of the following peaks; 2θ = 25.32°, 36.96°, 37.78°, 48.08°, 53.9°, 55.05°, 62.68°, 68.77°, 70.26°, and 75.05°. These reflections correspond to Miller's indices of (101), (103), (004), (200), (105), (211), (204), (116), (220), and (215), respectively, as mentioned above their XRD peaks. This result confirms the formation of anatase TiO_2 of lattice parameter a = 3.21 Å, which is consistent with the JCPDS card no. 21–1272. In addition, the inset of this figure shows an FE-SEM image of the sol-gel prepared TiO_2 that shows nano-particle morphology with an average particle size of 31.52 nm. This result is consistent with the estimated crystallite size by XRD. **Figure 15(d)** shows the XRD spectrum of the sol-gel/hydrothermally prepared $NaTiO_3$. The peaks at 8.71°, 24.25°, 28.47°, and 48.37° are consistent with the JCPDS card no. 47–0124, confirming the high purity of the material [49]. The inset of this figure shows the HR-TEM image. $NaTiO_3$ displays nanofibers-like morphology, with diameters in the range of 7.7–21.5 nm and lengths up to ~79 nm.

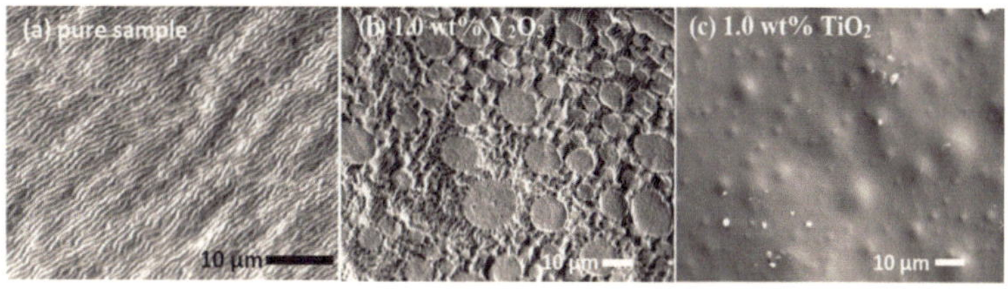

Figure 16.
(a-c) SEM images for PVAc/PMMA blended and the blend loaded with Y_2O_3 nanosheets and TiO_2 nanoparticles.

When Y_2O_3 or TiO_2 are introduced to PVAc/PMMA blend to make polymer nanocomposites, the morphology of this sol-gel prepared nanopowder did not change. **Figure 16(a–c)** shows the SEM image of PVAc/PMMA blend loaded with 1.0 wt% nanofillers. The un-doped blend surface exhibits a networked structure or wavy-like like the pool surface, **Figure 16(a)**. The fillers are distributed homogeneously and maintain their morphology, where Y_2O_3 distribute as nanosheets, **Figure 16(b)**, while TiO_2 is like small spheres of agglomerated particles, **Figure 16(c)**.

Moreover, differently prepared nanofillers were prepared to load with polymers and to get nanocomposites for suitable applications [24, 53–55]. The dielectric permittivity of poly(methyl methacrylate, PMMA, significantly increased while the dielectric loss remained almost low due to the doping with CuO/Co_3O_4 nanoparticles [53]. The semiconducting properties of poly(vinyl acetate)/poly(methyl methacrylate), P(VAc/MMA), were enhanced by adding TiO_2 nanoparticles and Y_2O_3 nanosheets to be used for some device applications [54]. The $Zn_{0.95}N_{0.05}O$ (ZNO) nanoparticles loaded with polyvinyl chloride (PVC) affected the optical as well as the dielectric properties of the pristine sample [55]. **Figure 17** represents the pronounced change in the absorbance of the polymeric materials by adding the ZNO and TiO_2 nanoparticles that were prepared using the sol-gel method.

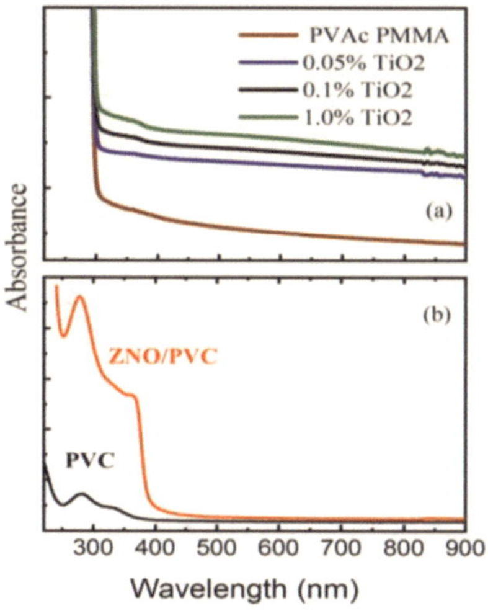

Figure 17.
The absorbance of PVAc/PMMA) loaded with TiO_2 nanoparticles (a), and PVC loaded with 5 wt.% ZNO nanoparticles (b).

7. Limitations exist for sol-Gel processing

- Sol-gel method process needs special attention at times of drying and aging.
- The reduction in the volume through the densification may cause shrinkage and cracking of the sample surface.
- Precursors used may be sensitive to moisture, and thus reduce the possibility of using the sol-gel materials on a large scale in optical coatings.
- The involved chemical reactions may yield undesired byproducts, which could affect the material's properties.
- The grade of the using precurses should be analytical in type to get pure materials.

8. Conclusion and outlook

The simplicity of the synthesizing process makes the sol-gel one of the most popular options in the coating industry and is projected to have a wide range of applications, with continued expansion. Recent progress in several applications was made by using the sol-gel technology, including the anti-reflection for solar cells, coating protection of aircraft, and cotton fabrics for flame retardant. In addition, the sol-gel process offers the possibility of large-area deposition, compared to vacuum-based-deposition processes, with the possible control of the microstructure and density of the films for improving the coloring efficiency and storage capacity of electrochromic films. Other advantages can be listed as:

- The low-temperature applied in all stages, except the densification, inhibits the thermal degradation of the material. Also, the development of amorphous, porous, and nanocrystalline materials is easily possible.
- Most of the precursors, solvents, and chelating agents are volatile, so the obtained materials are of high purity.
- Using the miscible precursors permits a homogeneous control over doping.
- The synthesis procedure requires only kind of chemical conditions.
- The easy casting of the materials into different shapes, such as monoliths, films, fibers, etc. However,

    1. More research on the effect of several preparative parameters in the sol-gel synthesis on the final structure of products and their physicochemical properties should be carried out. These parameters (solution molarities and concentration, pH, temperature, aging and preannealing/annealing times, and drying conditions).
    2. Utilizing the sol-gel technology for CO_2 capture within the developed materials (carbon-capture materials) will be a promising research area for a safe environment.

Acknowledgements

This work is funded by the Deputyship of Research & Innovation, Ministry of Education in Saudi Arabia, through project number) 904/1443). In addition, the authors would like to express their appreciation for the support provided by the Islamic University of Madinah.

Declaration (conflict of interest)

The authors declare that they have no known competing financial interests or personal relationships that could have appeared to influence the work reported in this paper. Therefore, there are no interests to declare toward any financial interests/ personal relationships that may be considered potential competing interests.

Author details

Arafa Hassen[1*], Adel M. El Sayed[1*], Azza Al-Ghamdi[2,3] and Mohamed Shaban[4,5]

1 Faculty of Science, Physics Department, Fayoum University, El-Fayoum, Egypt

2 Department of Chemistry, College of Science, Imam Abdulrahman Bin Faisal University, Dammam, Saudi Arabia

3 Basic and Applied Scientific Research Center (BASRC), Renewable and Sustainable Energy Unit, Imam Abdulrahman Bin Faisal University, Dammam, Saudi Arabia

4 Faculty of Science, Department of Physics, Islamic University of Madinah, Al-Madinah Al-Munawarah, Saudi Arabia

5 Faculty of Science, Physics Department, Nanophotonics and Applications (NPA) Lab, Beni-Suef University, Beni-Suef, Egypt

*Address all correspondence to: ash02@fayoum.edu.eg and ams06@fayoum.edu.eg

References

[1] Klabunde KJ, Stark J, Koper O, Mohs C, Park DG, Decker S, et al. Nanocrystals as stoichiometric reagents with unique surface chemistry. Journal of Physical Chemistry. 1996;**100**(30): 12142-12153. DOI: 10.1021/jp960224x

[2] Nair PAK, Vasconcelos WL, Paine K, Holley JC. A review on applications of sol-gel science in cement. Construction and Building Materials. 2021;**291**:123065. DOI: 10.1016/j.conbuildmat.2021.123065

[3] Zanurin A, Johari NA, Alias J, Ayu HM, Redzuan N, Izman S. Research progress of sol-gel ceramic coating: A review. Materials Today: Proceedings. 2022;**48**:1849-1854. DOI: 10.1016/j.matpr.2021.09.203

[4] Ozer N, Lampert CM. Electrochromic characterization of sol-gel deposited coatings. Solar Energy Materials and Solar Cells. 1998;**54**:147-156. DOI: 10.1016/S0927-0248(98)00065-8

[5] Borlaf M, Moreno R. Colloidal sol-gel: A powerful low-temperature aqueous synthesis route of nanosized powders and suspensions. Open Ceramics. 2021;**8**: 100200. DOI: 10.1016/j.oceram.2021.100200

[6] Simon SM, George G, Sajna MS, Prakashan VP, Jose TA, Vasudevan P, et al. Recent advancements in multifunctional applications of sol-gel derived polymer incorporated TiO_2-ZrO_2 composite coatings: A comprehensive review. Applied Surface ScienceAdvances. 2021;**6**:100173. DOI: 10.1016/j.apsadv.2021.100173

[7] Eltoum MSA, Nasr RMO, Omer HMA. Preparation and characterization of CuO Nano particles using sol-gel method and its application as CuO/Al_2O_3 supported catalyst. American Journal of Nano Research and Applications. 2020;**8**(2):16-21. DOI: 10.11648/j.nano.20200802.11

[8] Löbmann P. Antireflective coatings by sol-gel processing: Commercial products and future perspectives. Journal of Sol-Gel Science and Technology. 2017;**83**:291-295. DOI: 10.1007/s10971-017-4408-x

[9] Surekha P, Varshney AD, Jerusha E, Pant B, Rajesh AS. Optical applications of sol-gel nano-composites. Materials Today: Proceedings. 2022;**62**:2034-2037. DOI: 10.1016/j.matpr.2022.02.429

[10] Chen C, Yu M, Zhan Z, Ge Y, Sun Z, Liu J. Effect of pH on the structure and corrosion protection properties of sol-gel coatings. Corrosion Science. 2023;**212**: 110955. DOI: 10.1016/j.corsci.2022.110955

[11] Szołdra P, Frąc M, Pichór W. Effect of sol composition on the properties of TiO_2 powders obtained by the sol-gel method. Powder Technology. 2021;**387**: 261-269. DOI: 10.1016/j.powtec.2021.04.037

[12] Luo T, Luo C, Shi Z, Li X, Wu F, Zhang L. Optimization of sol-gel combustion synthesis for calcium looping CO_2 sorbents, part I: Effects of sol-gel preparation and combustion conditions. Separation and Purification Technology. 2022;**292**:121081. DOI: 10.1016/j.seppur.2022.121081

[13] Soria AC, SÚnchez JL, Miralles CG, Varela M, Navarro E, Gonalez C, et al. Novel one-pot sol-gel synthesis route of Fe_3C/few-layered graphene core/shell nanoparticles embedded in a carbon matrix. Journal of Alloys and Compounds. 2022;**902**:163662. DOI: 10.1016/j.jallcom.2022.163662

[14] Hashjin RR, Ranjbar Z, Yari H, Momen G. Tuning up sol-gel process to achieve highly durable superhydrophobic coating. Surfaces and Interfaces. 2022;**33**:102282. DOI: 10.1016/j.surfin.2022.102282

[15] Kesim MT, Durucan C. Indium tin oxide thin films elaborated by sol-gel routes: The effect of oxalic acid addition on optoelectronic properties. Thin Solid Films. 2013;**545**:56-63. DOI: 10.1016/j.tsf.2013.07.031

[16] El-Kemary M, Nagy N, El-Mehasseb I. Nickel oxide nanoparticles: Synthesis and spectral studies of interactions with glucose. Materials Science in Semiconductor Processing. 2013;**16**: 1747-1752. DOI: 10.1016/j.mssp.2013.05.018

[17] El Sayed AM. Opto-structural and surface properties of silkworm-like nickel oxide thin films. Materials Research Express. 2019;**6**:116423. DOI: 10.1088/2053-1591/ab4663

[18] El Sayed AM. Exploring the morphology, optical and electrical properties of nickel oxide thin films under lead and iridium doping. Physica B. 2021;**600**:412601. DOI: 10.1016/j.physb.2020.412601

[19] El Sayed AM, Mohamad ADM. Synthesis, structural, thermal, optical and dielectric properties of chitosan biopolymer; influence of PVP and α-Fe_2O_3 Nanorods. Journal of Polymer Research. 2018;**25**:175. DOI: 10.1007/s10965-018-1571-x

[20] El Sayed AM, Saber S. Structural, optical analysis, and Poole–Frenkel emission in NiO/CMC–PVP: Bio-nanocomposites for optoelectronic applications. Journal of Physics and Chemistry of Solids. 2022;**163**:110590. DOI: 10.1016/j.jpcs.2022.110590

[21] Colak H, Turkoglu O. Structural and electrical studies of Cu-doped CdO prepared by solid state reaction. Materials Science in Semiconductor Processing. 2013;**16**:712-717. DOI: 10.1016/j.mssp.2012.12.013

[22] Bazargan AM, Fateminia SMA, Ganji ME, Bahrevar MA. Electrospinning preparation and characterization of cadmium oxide nanofibers. Chemical Engineering Journal. 2009;**155**:523-527. DOI: 10.1016/j.cej.2009.08.004

[23] Tadjarodi A, Imani M. Synthesis and characterization of CdO nanocrystalline structure by mechanochemical method. Materials Letters. 2011;**65**:1025-1027. DOI: 10.1016/j.matlet.2010.12.054

[24] El Sayed AM, El-Sayed S, Morsi WM, Mahrous S, Hassen A. Synthesis, characterization, optical and dielectric properties of polyvinyl chloride/cadmium oxide nanocomposite films. Polymer Composites. 2014;**35**: 1842-1851. DOI: 10.1002/pc.22839

[25] Okabayashi J, Kono S, Yamada Y, Nomura K. Magnetic and electronic properties of Fe and Ni codoped SnO_2. Journal of Applied Physics. 2012;**112**: 073917. DOI: 10.1063/1.4754454

[26] Marikkannan M, Vishnukanthan V, Vijayshankar A, Mayandi J, Pearce JM. A novel synthesis of tin oxide thin films by the sol-gel process for optoelectronic applications. AIP Advances. 2015;**5**: 027122. DOI: 10.1063/1.4909542

[27] El Sayed AM, Morsi WM. α-Fe_2O_3/(PVA + PEG) nanocomposite films; synthesis, optical, and dielectric characterizations. Journal of Materials Science. 2014;**49**:5378-5387. DOI: 10.1007/s10853-014-8245-9

[28] El Fewaty NH, El Sayed AM, Hafez RS. Synthesis, structural and

optical properties of tin oxide nanoparticles and its CMC/PEG–PVA nanocomposite films. Polymer Science Series A. 2016;**58**:1004-1016. DOI: 10.1134/S0965545X16060055

[29] Li S, Yang W, Chen M, Gao J, Kang J, Qi Y. Preparation of PbO nanoparticles by microwave irradiation and their application to Pb(II)-selective electrode based on cellulose acetate. Materials Chemistry and Physics. 2005;**90**: 262-269. DOI: 10.1016/j.matchemphys. 2004.02.022

[30] Lu BT, Luo JL, Lu YC. Passivity degradation of nuclear steam generator tubing alloy induced by Pb contamination at high temperature. Journal of Nuclear Materials. 2012;**429**: 305-314. DOI: 10.1016/j.jnucmat. 2012.06.021

[31] Kumaravel R, Krishnakumar V, Ramamurthi K, Elangovan E, Thirumavalavan M. Deposition of $(CdO)_{1-x}(PbO)_x$ thin films by spray pyrolysis technique and their characterization. Thin Solid Films. 2007; **515**:4061-4065. DOI: 10.1016/j. tsf.2006.10.128

[32] Motlagh MMK, Mahmoudabad MK. Synthesis and characterization of lead oxide nano-powders by sol-gel method. Journal of Sol-Gel Science and Technology. 2011;**59**:106-110. DOI: 10.1007/s10971-011-2467-y

[33] El Sayed AM, Morsi WM. Dielectric relaxation and optical properties of polyvinyl chloride/Lead monoxide nanocomposites. Polymer Composites. 2013;**34**:2031-2039. DOI: 10.1002/ pc.22611

[34] Khataee A, Gholami P, Vahid B. Catalytic performance of hematite nanostructures prepared by N_2 glow discharge plasma in heterogeneous Fenton-like process for acid red 17 degradation. Journal of Industrial and Engineering Chemistry. 2017;**50**:86-95. DOI: 10.1016/j.jiec.2017.01.035

[35] Qayyum HA, Al-Kuhaili MF, Durrani SMA. Investigation of fundamental and high order optical transitions in α-Fe_2O_3 thin films using surface barrier electroreflectance. Superlattices and Microstructures. 2017; **110**:98-107. DOI: 10.1016/j.spmi. 2017.08.057

[36] Alqasem B, Yahya N, Qureshi S, Irfan M, Rehman ZU, Soleimani H. The enhancement of the magnetic properties of α-Fe_2O_3 nanocatalyst using an external magnetic field for the production of green ammonia. Materials Science and Engineering B. 2017;**217**: 49-62. DOI: 10.1016/j.mseb.2016.12.002

[37] Ramya SIS, Mahadevan CK. Preparation and structural, optical, magnetic, and electrical characterization of $Mn^{2+}/Co^{2+}/Cu^{2+}$ doped hematite nanocrystals. Journal of Solid State Chemistry. 2014;**211**:37-50. DOI: 10.1016/j.jssc.2013.11.022

[38] Zeng P, Zhao Y, Lin Y, Wang X, Li J, Wang W, et al. Enhancement of electrochemical performance by the oxygen vacancies in hematite as anode material for lithium-ion batteries. Nanoscale Research Letters. 2017;**12**:13. DOI: 10.1186/s11671-016-1783-0

[39] Tadić M, Čitaković N, Panjanc M, Stojanović Z, Marković D, Spasojević V. Synthesis, morphology, microstructure and magnetic properties of hematite submicron particles. Journal of Alloy Compounds. 2011;**509**:7639-7644. DOI: 10.1016/j.jallcom.2011.04.117

[40] Wang D, Wang Q, Wang T. Controlled synthesis of mesoporous hematite nanostructures and their

application as electrochemical capacitor electrodes. Nanotechnology. 2011;**22**: 135604. DOI: 10.1088/0957-4484/22/13/135604

[41] El Sayed AM. Influence of the preparative parameters on the microstructural, and some physical properties of hematite nanopowder. Material and Research Express. 2018;**5**: 025025. DOI: 10.1088/2053-1591/aaad36

[42] El-Sayed S, Hashhash A, Refai HS, Rutkauskas AV, Baleidy WS, Lis ON, et al. The detailed studies of the structural and magnetic properties of hexaferrites $Ba_{1-x}Sr_xFe_{12}O_{19}$ for $0.0 \leq x \leq 0.75$. Journal of Materials Science: Materials in Electronics. 2021;**32**:10977. DOI: 10.1007/s10854-021-05757-1

[43] Hashhasha A, Hassen A, Baleidy WS, Refaia HS. Impact of rare-earth ions on the physical properties of hexaferrites $Ba_{0.5}Sr_{0.5}RE_{0.6}Fe_{11.4}O_{19}$, (RE = La, Yb, Sm, Gd, Er, Eu, and Dy). Journal of Alloys and Compounds. 2021; **873**:159812. DOI: 10.1016/j.jallcom.2021.159812

[44] Khanjani S, Morsali A. Synthesis and characterization of lanthanum oxide nanoparticles from thermolysis of nanostructured supramolecular compound. Journal of Molecular Liquids. 2010;**153**:129-132. DOI: 10.1016/j.molliq.2010.01.010

[45] Whiffen RMK, Bregiroux D, Viana B. Nanostructured Y_2O_3 ceramics elaborated by spark plasma sintering of nanopowder synthesized by PEG assisted combustion method: The influence of precursor morphological characteristics. Ceramics International. 2017;**43**:15834-15841. DOI: 10.1016/j.ceramint.2017.08.153

[46] Bardak T, Tankut AN, Tankut N, Sozen E, Aydemir D. The effect of nano-TiO_2 and SiO_2 on bonding strength and structural properties of poly (vinyl acetate) composites. Measurement. 2016;**93**:80-85. DOI: 10.1016/j.measurement.2016.07.004

[47] Bsiri N, Zrir MA, Bardaoui A, Bouaïcha M. Morphological, structural and ellipsometric investigations of Cr doped TiO_2 thin films prepared by sol-gel and spin coating. Ceramics International. 2016;**42**:10599-10607. DOI: 10.1016/j.ceramint.2016.03.145

[48] Shi H, Magaye R, Castranova V, Zhao J. Titanium dioxide nanoparticles: A review of current toxicological data. Particle and Fibre Toxicology. 2013;**10**: 15. DOI: 10.1186/1743-8977-10-15

[49] El Sayed AM, El-Gamal S. Synthesis, optical, and electrical properties of starch/chitosan/$NaTiO_3$ bio-nanocomposites modified with $ErCl_3$. Physica Scripta. 2022;**97**:015805. DOI: 10.1088/1402-4896/ac40da

[50] El Sayed AM. Aspects of structural, optical properties, and relaxation in ($BiFeO_3$ or $NaTiO_3$)–PMMA: Hybrid films for dielectric applications. Journal of Physics and Chemistry of Solids. 2021; **148**:109767. DOI: 10.1016/j.jpcs.2020.109767

[51] Dal ND, Chavda NN, Madhad PH, Kumar R, Bhammar NA, Udeshi B, et al. Structural and electrical properties of pure and doped lanthanum oxide. International Journal of Modern Physics B. 2021;**35**(20):2150210. DOI: 10.1142/S0217979221502106

[52] Hajizadeh-Oghaz M, Razavi RS, Barekat M, Naderi M, Malekzadeh S, Rezazadeh M. Synthesis and characterization of Y_2O_3 nanoparticles by sol-gel process for transparent ceramics applications. Journal of Sol-Gel Science and Technology. 2016;**78**:

682-691. DOI: 10.1007/s10971-016-3986-3

[53] El-Sayed S, El Sayed AM. Preparation and characterization of CuO/Co_3O_4/poly(methyl methacrylate) nanocomposites for optical and dielectric applications. Journal of Materials Science: Materials in Electronics. 2021; **32**:13719. DOI: 10.1007/s10854-021-05949-9

[54] El-Sayed S, El Sayed AM. Influence of the sol-gel-derived Nano-sized TiO_2 and Y_2O_3 in improving the optical and electric properties of P(VAc/MMA). Brazilian Journal of Physics. 2021;**51**: 1584. DOI: 10.1007/s13538-021-00979-4

[55] Abdel-Baset T, El-Sayed S. The effect of $Zn_{0.95}Ni_{0.05}O$ nanoparticles on the physical properties of polyvinyl chloride. Polymer Bulletin. 2022;**79**:2915. DOI: 10.1007/s00289-021-03614-z

Section 3

Applications of Materials Based on Sol-Gel Synthesis

Chapter 5

Perspective Chapter: Mechanistic Understanding of Stability and Photocatalytic Efficiency of Titanium Dioxide Nanomaterials in Aquatic Media – A Sol-Gel Approach

Saptarshi Ghosh and Manoranjan Sahu

Abstract

Titanium dioxide (TiO_2) nanoparticles enhance the intrinsic value of commercial products like various cosmetics, paints, self-cleaning products, etc. Several research on the fabrication of TiO_2, stabilization of TiO_2 to retain its nanometric scale and increasing the inherent property of the material (i.e., photocatalytic) is ongoing for the last few decades. Still, the synthesis of highly efficient, stable, reproducible and cost-effective TiO_2 nanoparticles remains a grand challenge for the researchers and scientific community. Further research is needed to develop an in-depth understanding of synthesis, aggregation kinetics and efficiency to improve the performance of TiO_2 nanomaterial for the degradation of persistent organic pollutants (POPs). In this book chapter, we have summarized the synthesis process using the sol-gel pathway followed by its stability behavior and photocatalytic activity in the aqueous solutions. This study also highlighted the effects of various process variables such as pH, catalyst concentration, inorganic species etc. in the photocatalytic performance of TiO_2 nanoparticles. Finally, we have reviewed various strategies that have been performed for increasing the photocatalytic efficiency of TiO_2 by overcoming its limitations.

Keywords: doping, DLVO theory, photocatalysis, sol-gel process, TiO_2 nanoparticles

1. Introduction

Titanium is a transition metals having electronic configuration $3d^2 4S^2$. In nature, it is mainly available in the form of ilmenite ($FeTiO_3$) and binary metal oxides [1]. In a binary metal oxide form, Titania has three phases: anatase, brookite and rutile [2]. For the commercial purposes, anatase phase of titanium dioxide is largely used. It was found that, at high temperatures brookite and anatase phases transform to rutile, which is the most stable form of TiO_2. Now, as compared to bulk material, nanostructured TiO_2 has showed unique optical [3], electronic [4] and catalytic properties [5]. Due to these special properties, the interest for nanostructured titanium dioxide has grown to use it in various applications such as photocatalysis, paint pigments [6], cosmetics, personal care products [7], sensors, water and air purification, etc.

1.1 Processes for synthesis of nanomaterials

Nanomaterials synthesis processes can be broadly classified into top-down and bottom-up processes. In the bottom-up synthesis process, nanomaterials are synthesized from basic building units like, atoms or molecules. Whereas, in top-down approach, bulk materials are broken down into smaller counter parts. Top-down synthesis includes processes such as laser ablation, sputtering and grinding, ball milling, etc. In last few decades, use of nanomaterials have tremendously increased in several industries such as paint, cosmetics, catalyst, etc. In this regard, various synthesis process has been implemented and followed for fulfilling the industrial demand for nanomaterials. The commonly used nanomaterials synthesis processes are as follows:

a. Condensation process: it is a bottom-up synthesis process in which nanoparticles are synthesized through flame pyrolysis.

b. Chemical vapor deposition: a process in which thin film is deposited onto a substrate through chemical reactions of the gaseous precursors.

c. Mechanical milling: it is a top-down process where nanoparticles are synthesized by breaking down of larger molecules into smaller units of nanometer range by the effect of mechanical device called mill.

1.2 Sol gel process for TiO_2 synthesis

At laboratory scale, sol-gel is the most commonly used process for nanomaterials synthesis. It is a bottom-up process in which a colloidal suspension is formed by dispersing solid particles in a liquid called a Sol and agglomerate together to form a continuous three-dimensional network extending throughout the liquid called a gel. Generally, a sol-gel process is based on following steps [8]:

i. Hydrolytic polycondensation: it is the first step in a sol-gel process in which a sol is formed by the hydrolysis of alkoxide precursors.

ii. Gelation: at gel point, uniform suspension starts forming by deposition of substrate into the sol known as gel and the process is gelation.

iii. Drying: drying is performed to remove the liquids from the gel and a major change to the network structure may occur. The gel structure is maintained by forming aerogel in which air is replacing the liquid. Similarly, a xerogel is formed by normal drying in which gel structure collapses because capillary forces drawing the walls of the pores together and reducing the pore size. Cracking may occur when the tension in the gel is so large that it cannot shrink anymore.

iv. Calcination: it is a thermal treatment in which very high temperature is employed to crystallize the nanomaterials by removing organics and volatile substances.

2. Stability of nanomaterials

To retain the properties of nanomaterials, one of the biggest challenges is the agglomeration of nanoparticles. The stability of nanomaterials depends on two opposite forces; one being the electrostatic repulsive force and second is van der Waals forces of attraction. Deryaguin-Landau-Verwey-Overbeek (DLVO) theory first gave this concept of opposite forces in a suspension. According to DLVO theory, particle stability depends on particle to particle and surface to particle interactions but, it varies for nanomaterials of different structures and heterogeneity. The DLVO theory does not consider all these factors, which make the calculation of nanomaterials stability more challenging. The potential energy of interaction due to the Van Der Waals force is given by the expression

$$V_a = \frac{-A}{6}\left[\frac{2a^2}{S(S+4a)} + \frac{2a^2}{S(S+2a)^2} + \ln\frac{S(S+4a)}{S(S+2a)^2}\right]$$

where A is the Hamaker constant, a is the particle radius, and s is the separation distance between the particle surfaces [9]. The Hamaker constant indicates the strength of mutual attraction between two colloidal particles and depends on material properties, and was 4×10^{-20} J for TiO_2 [10]. Similarly, potential energy between particles due to electrostatic repulsion is expressed as follow

$$V_r = 2\pi\varepsilon_0\varepsilon_r a\zeta^2 \ln\left(1 + e^{-ks}\right)$$

where $\varepsilon_0\varepsilon_r$ is the dielectric permittivity of the medium, ζ is the potential at the surface of the particle, and k is the Debye length (reciprocal of double layer thickness). Now, the total potential energy is calculated as $V_t = V_\alpha + V_r$. Another drawback of DLVO theory is that it did not considered hydration and Born forces in calculation [9].

Now, when nanoparticles are entered into the environment, due to influence of various physicochemical factors and environmental conditions, they got agglomerated and with time these agglomerates might transformed into strong aggregates or rather loose agglomerates. Various studies reported that mobility of TiO_2 in porous media is affected by various factors such as pH of the medium, ionic strength, presence of Natural Organic matters (NOMs), etc. [11, 12]. It was observed that with rise in pH, the mobility of TiO_2 nanoparticles increases whereas ionic strength promotes the retention of nanoparticles in porous media. The reason for the retention at higher ionic concentration is because of enhanced agglomeration resulting into greater retention.

Similarly, Brownian motion is another important factor of colloids or nanoparticles for collisions with each other. The reasons for such collisions might be due to attractive forces or repulsive forces. In the colloidal systems, the dominant attractive force is Van Der Waals forces and to maintain the stability, an adequately strong repulsive force is required to neutralize the van der Waals attractions. In general, in a colloidal systems stability was achieved by the addition of surfactants [13], polymers [14] and/ or electrostatically.

2.1 Stability of TiO_2 suspensions

One of the major limitations of TiO_2 nanoparticles in the suspension was its agglomeration because it reduces the surface area of nanoparticles resulting into reduction in catalytic performance [15]. Thus, it is important to identify and study the role of different parameters which significantly contributes towards agglomeration of TiO_2 nanoparticles. Various parameters influence the stability of nanoparticles such as, ionic strength, pH, concentration of NPs, sticking coefficient, etc. But, predominantly it is the surface charge which controls the stability of nanoparticles. In particular, if the particles have a zeta potential more than ± 30 mV, they are considered stable [16]. However, addition of salts in a solution increases the ionic strength and promotes the attraction with opposite charge particles hence, their sticking efficiency increases. Basically, salt ions will reduce the thickness of electrostatic double layer and allows particles to come closer to each other by attractive forces and forms agglomerates [17]. In addition, collisions between particles also promotes aggregation and the Brownian motion is the driving force to bring collisions among nanoparticles in the aqueous medium. Another parameter affecting the stability of nanoparticles is the pH of the medium. Now, depending on the pH value, surface ionization happens and it controls the surface charge of metal oxides present in water, including TiO_2.

3. Titanium dioxide nanoparticles as photocatalyst

TiO_2 nanoparticles have been considered and studied widely as one of the most effective photocatalyst for the oxidation of organic pollutants [18]. It is because of its thermodynamic stability, nontoxicity, economically cheap and have higher photocatalytic performance comparing to other semiconductors [19]. The crystalline phases of TiO_2 are anatase, rutile and brookite. Among these three, the most active form of TiO_2 is anatase because of its high adsorptive rate (active sites), high hole trapping capacity and most thermally stable crystal phase of TiO_2. TiO_2 has an energy gap (E_{bg}) of 3.2 eV for anatase type and it performs both oxidation as well as reduction. However, TiO_2 is active only under UV region (λ < 390 nm) and the earth's solar spectrum have around 5% of that. Thus, it performs limited photocatalysis in the visible regions.

3.1 Mechanisms of photocatalysis

The photocatalysis by TiO_2 favors both oxidation (valence band) and reduction (conduction band) on its surface, typically in the band edges. In a TiO_2 or in other semiconductor, the highest occupied molecular orbital is separated from the lowest unoccupied molecular orbital by an energy band gap (E_{bg}). When the photocatalyst comes under the light irradiation that have the energy greater than or equal to the band gap, the photoexcitation takes place as illustrated in **Figure 1**. It transfers the electron from the valence band of a semiconductor to the conduction band of a semiconductor, leaving behind a hole in the valence band [20]. This ultimately forms a positively charged hole and an electron i.e., electron-hole pairs or exciton. These photogenerated charged pairs have a strong tendency to recombine, it may be a surface recombination or volume recombination. Thus, if we can reduce the recombination of these electron-hole pairs, it can create a chain of photochemical reactions as stated below [21]:

i. Photoexcitation: $TiO_2 + h\nu \rightarrow e^- + h^+$

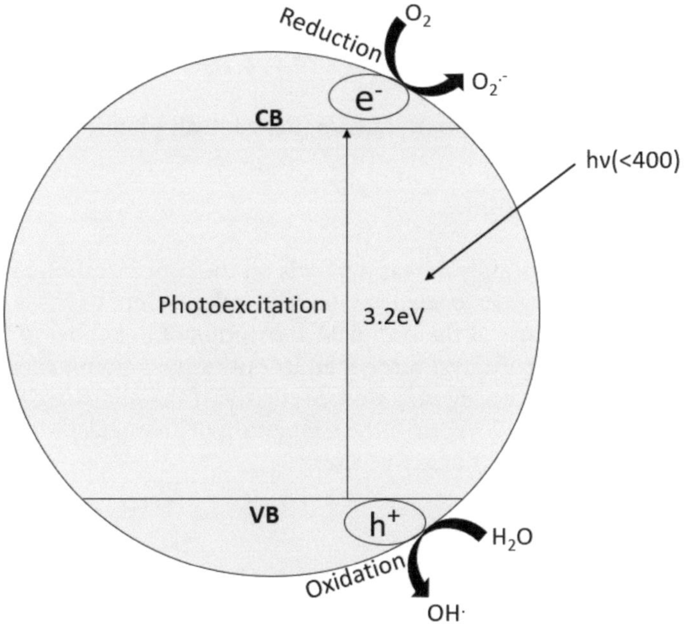

Figure 1.
Mechanisms of reactive oxygen species (ROS) generation by TiO_2 under illumination.

ii. Entrapment of free electrons: $e_{CB}^{-} \rightarrow e_{TR}^{-}$

iii. Entrapment of holes: $h_{VB}^{+} \rightarrow h_{TR}^{+}$

iv. Charge carrier recombination: $e_{TR}^{-} + h_{VB}^{+} \rightarrow e_{CB}^{-}$ + heat

v. Photoexcited electro scavenging: $(O_2)_{ads} + e^{-} \rightarrow O_2^{-}$

vi. Oxidation of hydroxyls: $OH^{-} + h^{+} \rightarrow OH^{\cdot}$

vii. Photodegradation by $OH^{\cdot}$ radicals: $R\text{-}H + OH^{\cdot} \rightarrow R'^{\cdot} + H_2O$

3.2 Factors influencing photocatalysis

3.2.1 Effect of pH

For the metal oxide like TiO_2 the surface charge may be positive or negative due to amphoteric behavior. Zero Point Charge (ZPC) represents the pH in which the catalyst does not have any charge on the surface. Various literature concluded that the ZPC of TiO_2 comes around 6–8 [22]. If the system is acidic (pH < pH ZPC), the catalyst protonated to gain positive charge on the surface and consecutively attracts negatively charged pollutants, but for alkaline conditions (pH > pH ZPC), the catalyst deprotonated to gain negative charge on the surface and favors the attraction of positively charged contaminants. The chemical transformation of TiO_2 with pH is as follows:

$$pH < pH\ ZPC : TiOH + H^+ \rightarrow TiOH_2^+ \text{ (protonation)}$$
$$pH > pH\ ZPC : TiOH + OH^- \rightarrow TiO^- + H_2O \text{ (deprotonation)}$$

Hence, pH is one of the important factors that controls photocatalysis.

3.2.2 Effect of catalyst concentration

The efficiency of photocatalysis was depends on the concentration of the photocatalyst used, but the performance rates was independent of the concentration above a certain limit because of the complete absorption of light energy by the photocatalyst [23]. Adding catalyst more than its optimum concentration causes excessive scattering of irradiating radiation. Similarly, if the photocatalyst concentration is less than its optimum level then the generation of charged species was inadequate due to lack of availability of active sites.

3.2.3 Effect of inorganic species

Presence of inorganic species can alter the photocatalytic process in various ways. Species like H_2O_2, $S_2O_8^{2-}$, etc. can improve the photocatalytic removal of pollutants by acting as electron trappers, thus enhancing the synthesis of $OH^·$ and other reactive oxygen species. But, the excessive loadings of inorganic species can reduce the photocatalytic process by taking up the $OH^·$ from the system. The photocatalytic efficiency of TiO_2 decreases with the availability of inorganic species like, $FeCl_2$, $Fe_2(SO_4)_3$, Na_2SO_4 and Na_2CO_3 because they alter the pH of the solution and also the active sites of the catalyst blocked by adsorption of ionic species. Ions that represent color in a solution can prevent the entry of light into the solution and decreases the photocatalytic process. Hence, ionic strength of a solution plays a prominent role in controlling the photocatalytic activity and pre-treatment was needed to achieve optimum photocatalytic treatment.

3.2.4 Effect of morphology of photocatalyst

Photocatalyst with various shape and size do not shows the exact photocatalytic behavior because of the dissimilarity in surface areas, charge transfer capacity and associated properties of the catalyst. It was reported that the photocatalytic performance is highly dependent on the morphologies and dimensions of the photocatalyst. The varied structure of the catalyst includes nanosphere, nanotubes, and nanosheets. All these structures show some peculiar properties that can enhance the photocatalysis. The zero-dimensional nanosphere have the high surface area where as the interconnected pore structure and shorter carrier-diffusion paths in three dimensional nanocubes and nanotubes helps in the movement of charged species. Hence, the morphology of the nanomaterial has a fascinating role in the photocatalytic processes and it is better to perform the experiment after modifying its shape. Future research work is needed to study the performance by changing the shape of the photocatalyst.

3.3 Modification of TiO_2 nanoparticles

TiO_2 has certain limitations and to improve the photocatalytic performance of TiO_2 various doping and nanocomposite strategies have been used. Recently, modifications of TiO_2 with carbon-based materials such as activated carbon, carbon nanotubes (CNT), [60]-fullerene and Graphene have received much attention to overcome the limitations of limited surface area. The mechanisms and strategies behind different doping patterns and nanocarbon-TiO_2 systems such as AC-TiO_2, CNT-TiO_2 and Graphene-TiO_2 nanocomposites are discussed in the below sections.

3.3.1 Doping

Doping means addition of impurities within the crystal lattice to extend its absorption to a longer wavelength. Doping may be cationic or anionic that can replace the atoms within the crystal lattice and creates structural defects. It also restricts the recombination of photo-induced charged carriers. Anionic dopants are considered to be useful for modification of TiO_2 because the p-orbitals of oxygen will interact with the p-orbitals of other anions and ultimately narrows down the band gap. Similarly, metal doping can stop the recombination of electron and hole pairs because the photo-excited charged species tends to move to the metals (electron traps). For cationic doping transition metals are generally preferred because they have a range of oxidation states. Thus, doping of TiO_2 by metal or non-metal brings reduction in the recombination rate of the excitons and promotes the absorption to longer wavelength. TiO_2 can be made visible light-active either by introducing structural defects such as Ti^{3+} and oxygen vacancies or by incorporation of non-metals such as N, C, and S, and transition metal ions. The various strategies of band gap modification are represented as follows:

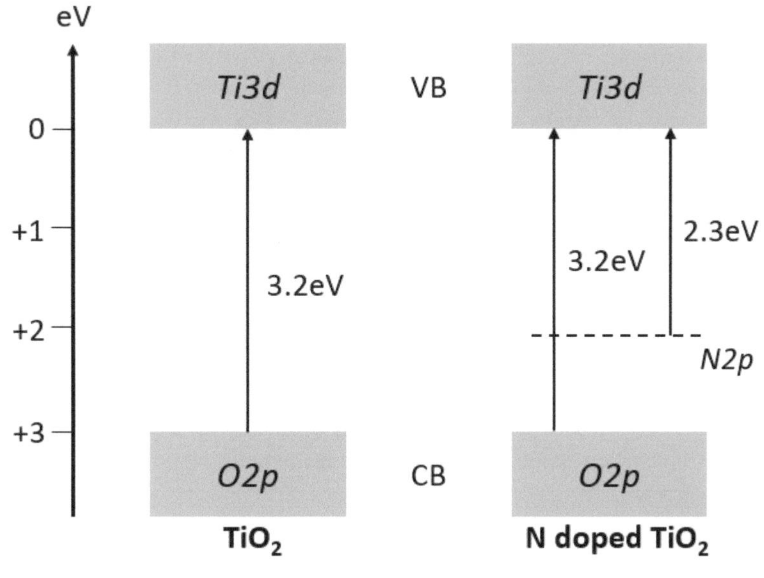

Figure 2.
Mechanism of band gap modification in nitrogen doped TiO_2. Modified and adapted from Ref. [24].

3.3.2 Formation of localized energy states

Doping with non-metal, such as N, C, F, S and B extends the light adsorption of TiO_2 towards the visible light region and thus enhances the photocatalytic activity. Among all the non-metal doped TiO_2, nitrogen doped TiO_2 has been found to exhibit superior photocatalytic activity under visible light irradiation. As shown in **Figure 2**, the ionic size of nitrogen was quite similar with the size of oxygen so that the orbital interactions (O2p orbital and N2p orbital) happens smoothly. And as the energy level for nitrogen was less than the oxygen, it forms a localized dopant state above the valence band of TiO_2 and helps in the shifting of the absorption range from UV to visible [24].

3.3.3 Broadening of valence band

The broadening of valence band mainly happens through the co-doping. TiO_2 co-doped with different elements, has attracted tremendous attention in the field of photocatalysis. Some studies reported that TiO_2 co-doped with appropriate elements could exhibit a much higher photocatalytic activity than any singly doped photocatalyst, because of the existence of synergistic effects between the doping elements. As soon as two different doping elements will be present in the TiO_2 matrix, their respective 2p orbital energy levels will both contribute to the creation of new energy states within the TiO_2 band gap in a synergetic manner. As compared to other non-metal elements, the co-doping of titanium dioxide with carbon and nitrogen (**Figure 3**) has been found to exhibit synergistic effect in a most pronounced manner resulting in a considerable visible light response of the respective co-doped materials [25].

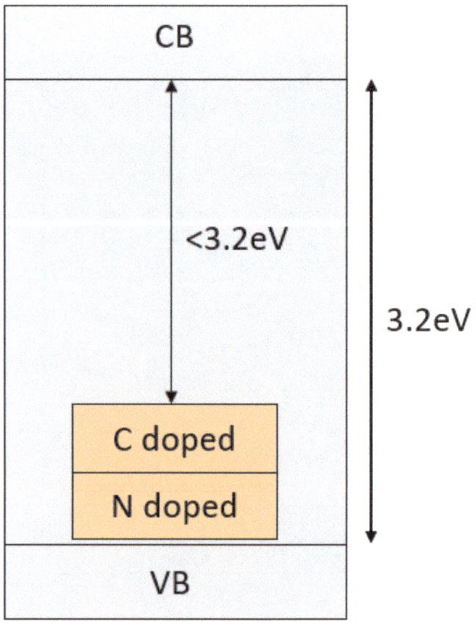

Figure 3.
Formation of new energy bands leading to thickening of VB in C—N co-doped TiO_2 Photocatalysts. Modified and adapted from Ref. [25].

3.3.4 Formation of color centres

Color centres give rise to formation of intra band gap energy states due to partial reduction of TiO_2. Color centres forms due to oxygen vacancy as well as electron vacancy in a metal oxide like TiO_2. Due to the loss of an O atom in a metal oxide the electron pair that remains trapped in the cavity V_O (reaction 1) left behind gives rise to an F centre (reaction 2), whereas a positively charged F^+ centre is equivalent to a single electron associated with the O vacancy (reaction 3). The electron-pair deficient oxygen vacancy, also known as an anion vacancy V_A, is referred to as a doubly charged F^{++} centre (reaction 4). Hence, the Color centres associated with oxygen vacancies are the F, F^+, and F^{++} centres in TiO_2. The electrons left in the V_O cavity can also interact with adjacent Ti^{4+} ions to give Ti^{3+} centres (reaction 5).

$$O^{2-} \rightarrow O + V_o + 2^{e-} \tag{1}$$

$$V_o + 2^{e-} \rightarrow F \tag{2}$$

$$V_o + e^- \rightarrow F^+ \tag{3}$$

$$V_o \rightarrow \{V_A \text{ or } F^{++}\} \tag{4}$$

$$F\ (\text{or } F^+) + Ti^{4+} \rightarrow F^+\ (\text{or } F^{++}) + Ti^{3+} \tag{5}$$

And, this Ti^{3+} (d^1) centres has a tendency to create a distortion within the crystal lattice called Jahn-Teller distortion that brings the electronic transition ($^2T_2 \rightarrow {}^2E$) in the crystal lattice as demonstrated in **Figure 4** and shifts the absorption into visible range [26].

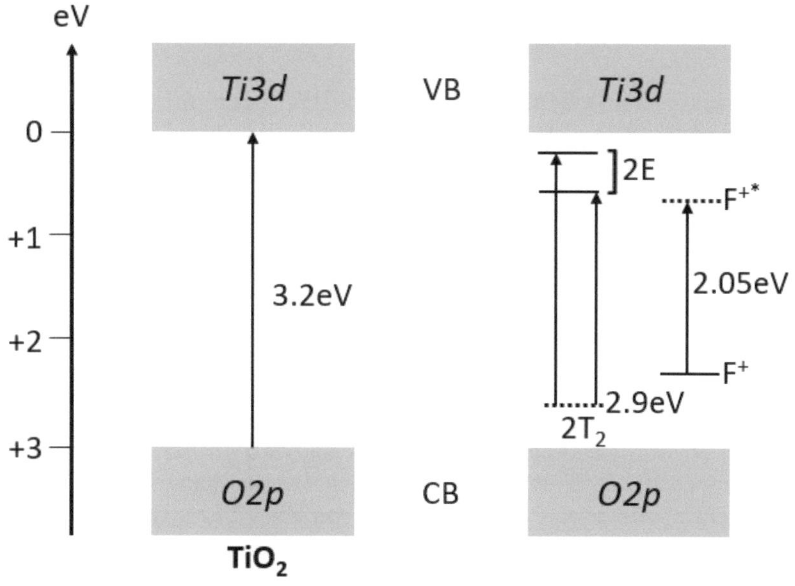

Figure 4.
Formation of color centre in TiO_2. Modified and adapted from Ref. [26].

3.3.5 Nanocomposite

TiO_2 based photocatalysis is one of the important catalytic process that can mineralizes the organic pollutant as well as the biological pollutant. But it suffers from low efficiency due to: (i) wide band gap (only activated under UV spectrum), (ii) very high recombination rate of photogenerated charged pairs, (iii) less surface area for the adsorption of pollutants. Now, the idea behind bringing composite of carbonaceous nanomaterials for improving the TiO_2 photocatalysis was to solve the limitations of TiO_2 and to promote forward reaction and adsorption of reactants through provision of adequate quality and quantity of active sites [27]. Various types of carbonaceous nanomaterials that are studied to make photocatalytic support (TiO_2-nanocarbon composite) are activated carbon, carbon nanotubes (CNT), and most importantly graphene which are discussed below.

3.3.6 Activated carbon-TiO_2 composites

Activated carbon is highly used because it is cheap, inert and easy to manufacture. Activated carbon possesses a porous amorphous structure with porosity spanning macro (>25 nm), meso (1–25 nm) and micro (<1 nm) pore ranges. The fundamental benefit of activated carbon is to provide a high surface area (typically 900–1200 m^2 g^{-1}) over which TiO_2 particles may be distributed and immobilized. The activated carbon support on TiO_2 particles increases the photocatalytic effect by synergistic effect. The synergistically enhanced photocatalytic activity was due to the adsorption of reactants on activated carbon followed by mass transfer to photoactive TiO_2 through the common interface between two. Activated carbon does not chemically interact with TiO_2, unless additional constituents are added to the system. So basically, activated carbon improves the photocatalytic process by adsorbing pollutants [28].

3.3.7 Carbon nanotubes (CNT)-TiO_2 composites

Carbon nanotubes may be single walled (SWCNTs) or multi walled (MWCNTs) having specific surface area of 400–900 $m^2\,g^{-1}$ and 200–400 $m^2\,g^{-1}$ respectively [28]. In addition to the high surface area, the surface chemistry of CNTs may be functionalized (incorporation of alcoholic group, ketonic group, carboxyl group, etc.) to promote specificity towards target pollutant. This represents an advantage over activated carbon, which are basically non-selective and therefore have a lower pollutant degradation rate due to breakdown of both target and neutral species. CNTs are widely reported to synergistically enhance the photocatalytic performance of TiO_2 through the retardation of electron-hole recombination as follows:

I. CNTs are capable of forming Schottky barriers at the CNT-TiO_2 interface, where there is a space charge region. And, TiO_2 is a n-type semiconductor but in the presence of a CNT, the photogenerated electron may move freely towards CNT surface, which may have a lower Fermi level. Thus, CNTs act as an electron sink and the excess holes in the Valence band of TiO_2 migrate to the surface and reacts. The TiO_2 now behaving as a P-type semiconductor. The high electrical conductivity and electron storage capacity (one electron for every 32 carbon atoms) makes CNT as highly effective electron sinks [28].

II. CNTs may enhance the TiO_2 activity by acting as a photosensitizer, transferring electrons to the TiO_2. This may be the reason behind the extending TiO_2 photocatalytic activity to a longer wavelength. In this mechanism the photogenerated electrons in the CNTs is transferred into the conduction band of TiO_2, promoting a reduction process (formation of superoxide anion by the adsorbed molecular oxygen). The positively charged CNTs then remove an electron from the valence band of TiO_2, leaving a hole. The positively charged TiO_2 then takes part in oxidation process with organic pollutants and water to form hydroxyl radicals.

3.3.8 Graphene-TiO_2 composites

Graphene is a single layered (one atom thick or thin) with a 2-D hexagonal structure arranged in a honeycomb lattice [29]. The sp^2 hybridized carbon atoms have delocalized π-bonds that improves the structural stability and current conductivity. Graphene has high electron mobility, acts as an adsorbent and promotes ballistic transport, meaning that electrons can travel without scattering, making them electron sinks or ideal electron transfer bridges. These unique properties of the graphene plane are mainly due to its long polyaromatic π-electron systems and coordinative unsaturated terminal carbon atoms. It has an energy band gap of 0 eV because it's filled valence bands touches the empty conduction bands in a certain point called Dirac point, thus giving rise to a new band [30]. Most importantly, graphene is chemically inert and highly stable under conditions where maximum substances would undergo rapid phase transformation and chemical reactions [29, 31, 32].

It was reported that graphene and its derivatives have been successfully used in designing nanocomposite systems for their excellent performance in photocatalysis. These functional nanomaterials have been used in designing photocatalyst systems because of their excellent electron transfer properties, which help to prevent recombination of photoexcited charge carriers in metal oxides (TiO_2). These also help in establishing good interaction of target molecules to the surface of the TiO_2 nanomaterials due to their adsorption behavior [33]. In summary, adsorption, conductivity, tuning of band gaps and very large planar structure of graphene plays a critical role for photodegradation of various kinds of pollutants over graphene-TiO_2 composites.

4. Conclusions

This book chapter covered various nanomaterials synthesis processes specifically sol-gel process, stability of nanomaterials in suspension and the photocatalytic process. Various parameters affecting the photocatalytic efficiency of TiO_2 in an aqueous medium is discussed. In addition, the chapter discusses various types of strategies that have been followed to improve the photocatalytic efficiency of TiO_2 and also highlighted the current strategies that have been performed. The book chapter clearly indicates that due to distinctive properties like, optical, electronic and catalytic, nanostructured TiO_2 becomes fascinating material for research. It is widely used in various environmental applications like disinfection, catalysis, sensing, etc. and for that extensively used in several industries. Thus, it is important to develop cost effective synthesis processes to support the industrial demands. Additionally, TiO_2

suffers from certain limitations such as aggregation, wide bandgap, recombination of excitons, etc. which need to be addressed through research and development of technology to enhance its catalytic efficiency so that it can resolve emerging environmental problems.

Acknowledgements

We are grateful to the University Grants Commission (UGC) and Indian Institute of Technology Bombay (IITB) for their support for this work.

Conflict of interest

The authors declare no conflict of interest.

Author details

Saptarshi Ghosh[1] and Manoranjan Sahu[1,2,3*]

1 Aerosol and Nanoparticle Technology Laboratory, Environmental Science and Engineering Department, Indian Institute of Technology Bombay, Mumbai, India

2 Inter-Disciplinary Program in Climate Studies, Indian Institute of Technology Bombay, Mumbai, India

3 Centre for Machine Intelligence and Data Science, Indian Institute of Technology Bombay, Mumbai, India

*Address all correspondence to: mrsahu@iitb.ac.in

References

[1] Shah KW, Huseien GF, Xiong T. Functional nanomaterilas and their applications toward smart and green buildings. In: New Materials in Civil Engineering. Butterworth-Heinemann; 2020. pp. 395-433

[2] Liao Y, Que W, Jia Q, He Y, Zhang J, Zhong P. Controllable synthesis of brookite/anatase/rutile TiO_2 nanocomposites and single-crystalline rutile nanorods array. Journal of Materials Chemistry. 2012;**22**(16): 7937-7944

[3] Zhu T, Gao SP. The stability, electronic structure, and optical property of TiO_2 polymorphs. The Journal of Physical Chemistry C. 2014;**118**(21): 11385-11396

[4] Aronne A, Fantauzzi M, Imparato C, Atzei D, De Stefano L, D'Errico G, et al. Electronic properties of TiO_2-based materials characterized by high Ti^{3+} self-doping and low recombination rate of electron-hole pairs. RSC Advances. 2017; 7(4):2373-2381

[5] Ochiai T, Fujishima A. Photoelectrochemical properties of TiO_2 photocatalyst and its applications for environmental purifications. Journal of Photochemistry and Photobiology C: Photochemistry Reviews. 2012;**13**(4): 247-262

[6] Middlemas S, Fang ZZ, Fan P. A new method for production of titanium dioxide pigment. Hydrometallurgy. 2013;**131**:107-113

[7] Serpone N, Dondi D, Albini A. Inorganic and organic UV filters: Their role and efficacy in sunscreens and suncare products. Inorganica Chimica Acta. 2007;**360**(3):794-802

[8] Fendler JH. Nanoparticles and Nanostructured Films: Preparation, Characterization, and Applications. John Wiley & Sons; 2008

[9] García-García S, Jonsson M, Wold S. Temperature effect on the stability of bentonite colloids in water. Journal of Colloid and Interface Science. 2006; **298**(2):694-705

[10] Zhang J. Surface forces between silica surfaces in CnTACl solutions and surface free energy characterization of talc [doctoral dissertation]. Virginia Tech; 2006

[11] Chen G, Liu X, Su C. Distinct effects of humic acid on transport and retention of TiO_2 rutile nanoparticles in saturated sand columns. Environmental Science & Technology. 2012;**46**(13):7142-7150

[12] Chowdhury I, Cwiertny DM, Walker SL. Combined factors influencing the aggregation and deposition of nano-TiO_2 in the presence of humic acid and bacteria. Environmental Science & Technology. 2012;**46**(13):6968-6976

[13] Wu N, Wei H, Zhang L. Efficient removal of heavy metal ions with biopolymer template synthesized mesoporous titania beads of hundreds of micrometers size. Environmental Science & Technology. 2012;**46**(1): 419-425

[14] Ji Z, Jin X, George S, Xia T, Meng H, Wang X, et al. Dispersion and stability optimization of TiO_2 nanoparticles in cell culture media. Environmental Science & Technology. 2010;**44**(19):7309-7314

[15] Ghosh S, Sahu M. Phthalate pollution and remediation strategies: A review.

Journal of Hazardous Materials Advances. 2022;**19**:100065

[16] Pate K, Safier P. Chemical metrology methods for CMP quality. In: Advances in Chemical Mechanical Planarization (CMP) 2022. Woodhead Publishing; 2022. pp. 355-383

[17] Jiang J, Oberdörster G, Biswas P. Characterization of size, surface charge, and agglomeration state of nanoparticle dispersions for toxicological studies. Journal of Nanoparticle Research. 2009; **11**(1):77-89

[18] Pang X, Skillen N, Gunaratne N, Rooney DW, Robertson PK. Removal of phthalates from aqueous solution by semiconductor photocatalysis: A review. Journal of Hazardous Materials. 2021; **402**:123461

[19] Bhanvase BA, Shende TP, Sonawane SH. A review on graphene-TiO_2 and doped graphene-TiO_2 nanocomposite photocatalyst for water and wastewater treatment. Environmental Technology Reviews. 2017;**6**(1):1-4

[20] Ahmed SN, Haider W. Heterogeneous photocatalysis and its potential applications in water and wastewater treatment: A review. Nanotechnology. 2018;**29**(34): 342001

[21] Nakata K, Fujishima A. TiO_2 photocatalysis: Design and applications. Journal of Photochemistry and Photobiology C: Photochemistry Reviews. 2012;**13**(3):169-189

[22] Ahmadi M, Motlagh HR, Jaafarzadeh N, Mostoufi A, Saeedi R, Barzegar G, et al. Enhanced photocatalytic degradation of tetracycline and real pharmaceutical wastewater using MWCNT/TiO_2 nano-composite. Journal of Environmental Management. 2017;**186**(186):55-63

[23] Herrmann JM. Heterogeneous photocatalysis: Fundamentals and applications to the removal of various types of aqueous pollutants. Catalysis Today. 1999;**53**(1):115-129

[24] Cheng X, Yu X, Xing Z. Characterization and mechanism analysis of N doped TiO_2 with visible light response and its enhanced visible activity. Applied Surface Science. 2012; **258**(7):3244-3248

[25] Schneider J, Matsuoka M, Takeuchi M, Zhang J, Horiuchi Y, Anpo M, et al. Understanding TiO_2 photocatalysis: Mechanisms and materials. Chemical Reviews. 2014; **114**(19):9919-9986

[26] Serpone N. Is the band gap of pristine TiO_2 narrowed by anion-and cation-doping of titanium dioxide in second-generation photocatalysts? The Journal of Physical Chemistry B. 2006; **110**(48):24287-24293

[27] Khalid NR, Majid A, Tahir MB, Niaz NA, Khalid S. Carbonaceous-TiO_2 nanomaterials for photocatalytic degradation of pollutants: A review. Ceramics International. 2017;**43**(17): 14552-14571

[28] Leary R, Westwood A. Carbonaceous nanomaterials for the enhancement of TiO_2 photocatalysis. Carbon. 2011;**49**(3):741-772

[29] Geim AK, Novoselov KS. The rise of graphene. Nature Materials. 2007;**6**(3): 183-191

[30] Bolotin KI, Sikes KJ, Jiang Z, Klima M, Fudenberg G, Hone J, et al. Ultrahigh electron mobility in suspended

graphene. Solid State Communications. 2008;**146**(9–10):351-355

[31] Min Y, Zhang K, Zhao W, Zheng F, Chen Y, Zhang Y. Enhanced chemical interaction between TiO_2 and graphene oxide for photocatalytic decolorization of methylene blue. Chemical Engineering Journal. 2012;**193**:203-210

[32] Ye H, Zhao B, Zhou Y, Du J, Huang M. Recent advances in adsorbents for the removal of phthalate esters from water: Material, modification, and application. Chemical Engineering Journal. 2021;**409**:128127

[33] Perreault F, De Faria AF, Elimelech M. Environmental applications of graphene-based nanomaterials. Chemical Society Reviews. 2015;**44**(16):5861-5896

Chapter 6

Perspective Chapter: Functional Sol-Gel Based Coatings for Innovative and Sustainable Applications

Silvia Sfameni, Giulia Rando and Maria Rosaria Plutino

Abstract

Alkoxysilanes represent a class of molecules widely employed to achieve the preparation of plenty of functional surfaces by easy, cost-effective and eco-friendly sol-gel methods. In this regard, the advancements of research activities include the proper design of film/patterns/brushes, by starting from opportune alkoxysilane and/or other metal/metalloid precursors, in order to obtain efficient innovative and homogenous functional surfaces showing implemented properties by means of the simple and eco-friendly sol-gel method. Therefore, in light of these aspects, the employment of opportune functional alkoxysilanes, either in combination with other nanofillers or molecules, is a key step for the design, and development of sol-gel based nanohybrid or nanocomposite coatings suitable for different surface properties implementation and applications, spanning from blue-growth sector to smart and technical textiles, from biomedicine to building and cultural heritages, from environmental remediation to catalysis. Some of the most relevant and explicative examples of these innovative and sustainable sol-gel based coatings will be described in this chapter.

Keywords: sol-gel, sustainable coating, protective finishing, blue-growth, cultural heritages, functional textiles

1. Introduction

One of the most recent trends in the field of nanotechnology is the development of hybrid functional nanomaterials and nanocomposites, characterized by the presence of functional nanometric organic or inorganic components appropriately dispersed in a (blended) polymeric matrix, thus allowing the fabrication of final hybrid materials with improved performances over that of the sum of the individual components [1–3]. In particular, these advanced nanostructured materials were often used as surface coatings for a wide range of substrates and therefore for different industrial application fields [4, 5]: among these, those related to the development of high-tech and smart surfaces are of great interest. Creating new functional coatings is crucial in order to give surfaces implemented properties like anti-vegetative [1], antibacterial [2], hydrophobic [3], anti-stain [4], and flame-retardant [5], and release of drugs [6], adsorption and detection of molecules [7]. In this regard, nowadays the sol-gel method could be unquestionably considered among the new, versatile, and sustainable scientific approaches that can lead to addressing the above-mentioned needs. Due to its benefits, such as its low processing temperature, increased homogeneity of the final products, and increased versatility of the corresponding silane precursors, as well as its ability to stably attach to molecules or a variety of substrates, the sol-gel method enables the creation of new functional materials using safe solvents (i.e. water, ethanol, etc.) and reactants with a look on the environment and human health [8].

1.1 Hybrid inorganic–organic materials

The manipulation of matter in the range of 1 to 100 nm is employed to design, engineer, and create nanomaterials in the emerging scientific and application field of nanotechnology.

Nanomaterials can be utilized for a wide range of applications and for developing different kinds of nanocomposites, nanohybrids and nano-devices because of their peculiar features mainly related to their nanoscale dimensions and morphology. As a matter of fact, the presence of their different physical, quantum, and surface features caused by small-scale confinement is the primary distinguishing factor between nanomaterials and the corresponding bulk materials, thus leading to a very interesting and useful wide range of useful physical, surface, and quantum characteristics [9]. Examples of these characteristics exploited at the nanometric scale can be represented by substances that can feature catalytic properties (i.e. gold and platinum), copper that turns transparent from opaque and silicon that turns into a conductor from an insulator [10]. From an application perspective, optical and electrical, chemical–physical, thermal, and mechanical characteristics of nanostructures are the most intriguing.

For example, as the name implies, a "nanohybrid", also often referred as Hybrid Inorganic–Organic Material (HIOM), points out any material that hosts a compound with nanometric dimensions bonded covalently within a polymeric matrix or structure, or that is produced by a supramolecular assembly of two stably interconnected nanomaterials with distinct former properties from the final formed hybrid itself [11]. More precisely, this wide class of inorganic–organic nanostructured materials can be classified into the following distinguished categories (**Figure 1**) [12]:

- composites, that is a material combination made up of a micrometric matrix and a dispersion;
- nanocomposites, which are sub-micrometre-sized mixtures of similar-natured materials (1–100 nm);
- hybrids, that is a sub-micrometric amalgamation of substances with a different nature from compound hybrid materials;
- nanohybrids, that is atomic or molecular mixtures of several materials held together by chemical bonds.

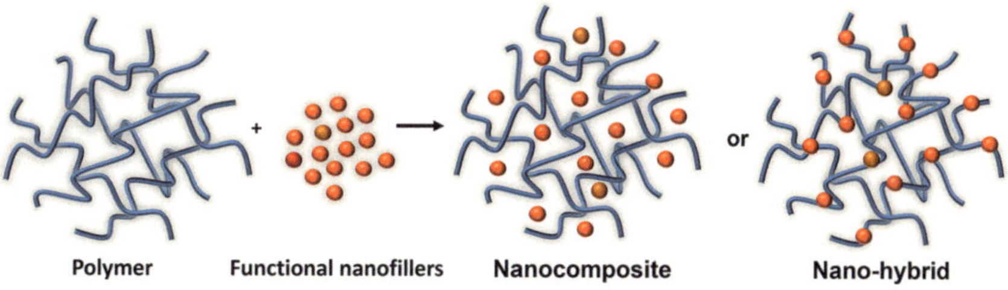

Figure 1.
Possibilities of nanocomposites and nanohybrid materials development by the addition of nanometric functional additives into a polymeric matrix.

Moreover, hybrid materials can be classified into two types of categories based on the nature of the interface [13, 14]:

- Class I hybrids, both organic and inorganic components are mixed, and weak connections bond the molecules in the overall structure (ionic bond, hydrogen bond or van der Waals forces).
- Class II hybrids, in which the two phases are chemically strongly connected (covalent bonds or ionic-covalent bonds).

Low temperatures are necessary for the synthesis of inorganic–organic hybrids since organic components are typically thermally unstable. In this regard, the sol-gel is a quite versatile approach that enables it to meet this need [15].

Hybrid inorganic–organic materials not only offer an intriguing alternative to the production of new compounds and materials useful in several research fields by enhancing their characteristics and properties; they can also be used to create new products in a variety of industrial sectors, including the optical, electronic, mechanical, energy, environmental, biological, and medical sectors [8].

It has been a long time since published research studies in literature examined the properties of organic substances, such as polymers and/or inorganic materials (i.e. metals, ceramics and glasses) in order to develop suitable materials, such as fibers or coatings. The links between the structure and properties of these organic and inorganic materials have been studied and rationalized by means of novel analysis techniques and spectroscopic technologies [16, 17].

In addition to their unique thermal and mechanical qualities, polymers are often used to produce hybrid materials because of their chemical functionalization, chemical stability, biocompatibility, optical and electrical capabilities, and balance between hydrophobicity and hydrophilicity.

The composition, size, crystallinity and structure of the inorganic phase, which can take the form of a variety of compounds, such as silicates, transition metal oxides, metal phosphates, nano clays, and nanostructured metals, determine the mechanical and thermal stability of inorganic molecules, as well as the introduction of new functionalities [17]. In particular, the hybrid magnetic, optical, electrical and redox characteristics can all be enhanced by the inorganic component [18].

Different techniques can be used to produce such hybrid functional materials. For instance, the condensation between oligomers or polymers bearing metal alkoxides could be obtained either by starting from low molecular weight alkyl/aryl(alkoxysilanes) and by introducing an opportune additive (e.g. nanoparticles or other nanomaterials) into an already swollen organic matrix, or by impregnating inorganic gels with a polymeric solution, and by employing precursors of general formula R'-$Si(OR)_3$, where R' is a polymerizable functional group (for example an epoxide or amino group) [19, 20].

In addition, the design and development of (multi)functional eco-friendly nanohybrids and nanocomposites that may find a useful application in multidisciplinary transversal fields could considerably contribute to the enhancement of human daily life and well-being, in the interest of long-term sustainability and environmental protection.

1.2 (multi)functional sol-gel based coatings

In recent years, the field of materials engineering has been active in the development of novel functional sol-gel based coatings to alter the characteristics of external surfaces [21]. This type of process, which leads to the formation of organic–inorganic hybrid coatings with a nanocomposite structure, is capable of modifying the surface properties such as abrasion resistance [22], wettability [23], as well as imparting antibacterial characteristics [24], resistance to UV radiation [20] and solvents in general and to release bioactive substances [25]. The sol-gel method is frequently used because of its versatility in the synthesis of inorganic–organic hybrid materials with a wide variety of mechanical, chemical, and morphological characteristics (**Figure 2**).

Alkoxides, chlorides of metals or metalloids are typical precursors and they take part in hydrolysis and condensation reactions to create sols, which are colloidal solutions of solid particles with sizes ranging from 1 nm to 1 μm in a liquid phase.

Then the sol phase develops until a continuous inorganic lattice, with an interconnected liquid phase (known as a gel), is formed [27]. Heat treatment is then frequently applied to stabilize the system, by eliminating the liquid phase from the gel and thus enhancing its mechanical characteristics. The sol phase can be processed in a variety of ways to create products with the desired shape; it can be poured into a mold to create massive products (monoliths, membranes, aerogels), used to create powders (nano and microspheres that can be used as finished products or as raw materials for the creation of sintered specimens), and finally, deposited on a surface to create thin or thick films.

Metallic or metalloid elements are often employed as precursors in the sol-gel process to create the colloid within various added ligands that may be bonded; among them, silane alkoxides are examples. Since the early 1800s, these latter are commonly used, due to their well-known chemistry, as well as the easiness with which they may react with water in the so-called hydrolysis reaction step. The silane alkoxides have a general structure of $Si(OR)_4$, where R is a variable organic functional group

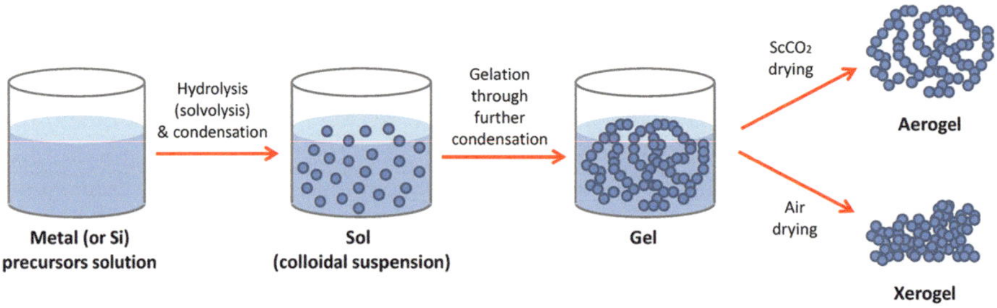

Figure 2.
Schematization of the sol–gel process. Reproduced from MDPI ref. [26].

depending on the alkoxide in question. Additionally, the inexpensive cost of these precursors makes it possible to use them in larger-industrial scale applications [28].

An initial hydrolysis reaction of the alkoxy groups, followed by a condensation event that creates a cross-linked system, is how a sol-gel system is created. As a result of these overall reaction steps, very large polymeric matrices with -Si-O-Si bridging bonds are produced. The flexibility of the final lattice phase decreases as the number of -Si-O-Si bonds rises, which causes the viscosity of the lattice to grow until it gels. The following variables can have a significant impact on the kinetics of formation of the individual reaction products and, consequently, on the final resulting one [29, 30], in particular:

i. the Rw ratio, also known as the degree of hydrolysis;

ii. the solvent;

iii. the addition of complexing compounds;

iv. the use of acidic or basic catalysts.

Thin sol-gel based films, often employed as paint, coating or finishing layers, have undoubtedly received the most attention in the industrial sector. Moreover, the versatility of the sol-gel method enables a rational design to control the properties of the final silane-based films by controlling its morphological structure at the nanometric scale and its adhesion to the substrate [31]. In addition, the sol-gel method allows the production of thin films using affordable, time-tested and favorable conditions (i.e. alcoholic or aqueous solvents, crosslinking at low temperatures). It is also possible to easily add more than one function to the same coating and therefore, obtain multi-functional coatings thanks to the potential of easily incorporating both organic and inorganic materials into the final sol-gel based formulation. There are several potential uses, formulations, and deposition techniques for the sol-gel materials that are currently used in the protective paints/coatings industry. While the design and development of opportune "sol-gel" necessitate significant expertise, the deposition to generate "xerogels" can be accomplished with straightforward and easy-to-use procedures for the end-users, such as:

- Spray coating [32];
- Dip-coating [33];
- Spin coating [34];
- The doctor-blading procedure [32].

Numerous industrial fields can benefit from the application of thin films made of silica, alumina, zirconium, titanium, or other inorganic oxides. For example, glass or plastic can feature anti-reflective properties and light alloys or polymers can have their hardness and wear resistance increased. The use of sol-gel coatings with an anticorrosion or anti-oxidation function is also fascinating. Hydrophilic glass coatings are appropriate for applications that call for anti-fog and antibacterial qualities. Other functionalities, such as hydrophilicity, hydrophobicity, or dirt-repellency properties, can be added by means of organic–inorganic topcoats [35].

The high degree of properties control and versatility of this method, which has a number of benefits over conventional techniques, as well as some drawbacks, piques curiosity [36–38].

Low process temperatures, good homogeneity, the ability to produce low-thickness films, the production of mixed oxides due to the stoichiometric control of the starting solution composition, better control of the material porosity by varying the heat treatment and a high degree of purity are all evident advantages.

Reversely, high initial material costs, potential fracture formation during the crosslinking step and lengthy process periods are some of the main disadvantages that need to be paid attention to.

2. Sol-gel based protective coatings for innovative and sustainable applications

2.1 Functional textiles

Due to current trends and client demands for high-tech or high-performance applications, functional nanostructured finishings for textile materials have received the most attention among the many finishing processes [39].

Since the coating process can affect a variety of significant textile properties, including comfort, breathability and the hand of the coated fabric, it is important to optimize the process. Furthermore, the surface of the fabric and the presence of impurities can both affect the coating adhesion to the textile fibers [40]. The main challenge relies on the achievement of both functional hi-tech and smart textiles with outstanding performance, uniformity and good coating distribution on the fabric surface. Textiles are typically prepared in accordance with specified treatments for these reasons.

In the last two decades, innovative applications to enhance the fundamental qualities of textiles have been developed with the help of sol-gel-assisted textile finishing approaches [41, 42]. Sol-gel based textile finishing typically offers greater benefits to make up for the drawbacks of traditional finishing methods (**Figure 3**).

The primary benefits include environmental friendliness, reduced chemical use, low-temperature processing, safety for human health, protection of the natural former qualities of textile materials and the ability to customize the coating thickness and the durability of completed fabrics [43]. In particular, the sol-gel method is a viable wet method for the deposition of functional hybrid organic–inorganic materials and nanoparticles, leading to a final functionalization of textiles. In comparison to other techniques, the integration of sol-gel technology into extensive industrial applications may be simpler and more convenient (**Figure 4**).

The widely used sol-gel coating application procedures are based on dip-coating, padding or spray, producing smart or functional textiles with precisely designed properties. From an economic standpoint, the dip-pad-cure method was shown to be the most popular due to its simplicity and viability [45]. Using a padder, the fabric is immersed or dipped into a coating material solution while moving at a constant speed. After drying and curing, the process is repeated. A layer with a thickness of about 100 nm is generated on the surface of the cloth throughout the drying and curing process (**Figure 4**). The functional groups of the precursors are thus attached to the surface of the textile to form covalent or stable hydrogen/ionic bonds, which greatly enhance the adhesion of the hybrid film to the textiles [44]. Through the

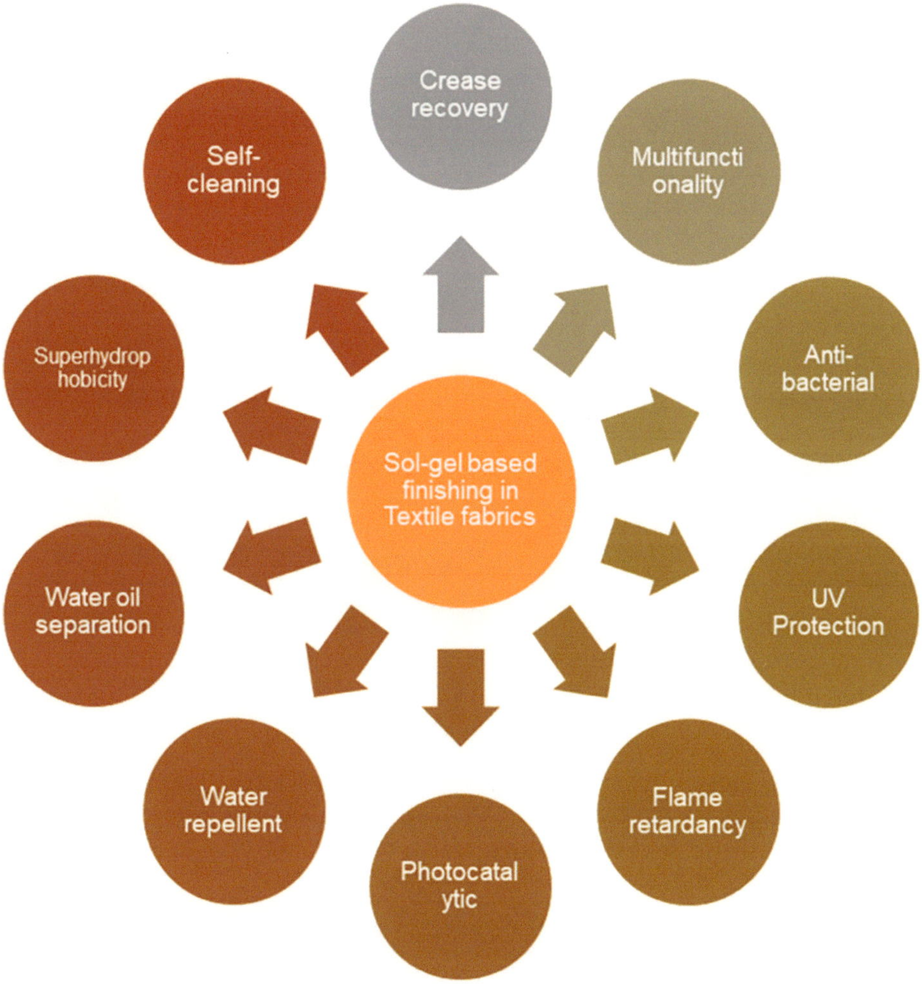

Figure 3.
Multifunctional properties achieved by the application of sol-gel based finishings for textile materials. Reproduced from MDPI ref. [41].

attaching of coating chemical moieties to the surface of the materials themselves, the sol-gel method, together with other process technologies, successfully modifies the surface of textile materials.

The recent last years have seen the application of hydrophobic treatment to textile surfaces for antifouling, self-cleaning, anti-ice and oil–water separation [46]. Additionally, due to their low surface energy and oil/water-repellent qualities, stain-resistant surfaces and anti-stain coatings have important applications in a variety of industries, including textiles, construction, cars and electronics [47–49]. For instance, the ability to repel water could aid in clearing bio-settlement off the surface and preventing textile damage. Specifically, because of the property of biomaterials, it immediately aids in the prevention of biofouling and bacterial colonization. Mechanical, chemical and coating functionalization techniques can all introduce these desired characteristics [50].

Water-repellence chemicals are added to the textile fiber to obtain the desired hydrophobic effect with less impact as possible on other functional intrinsic attributes

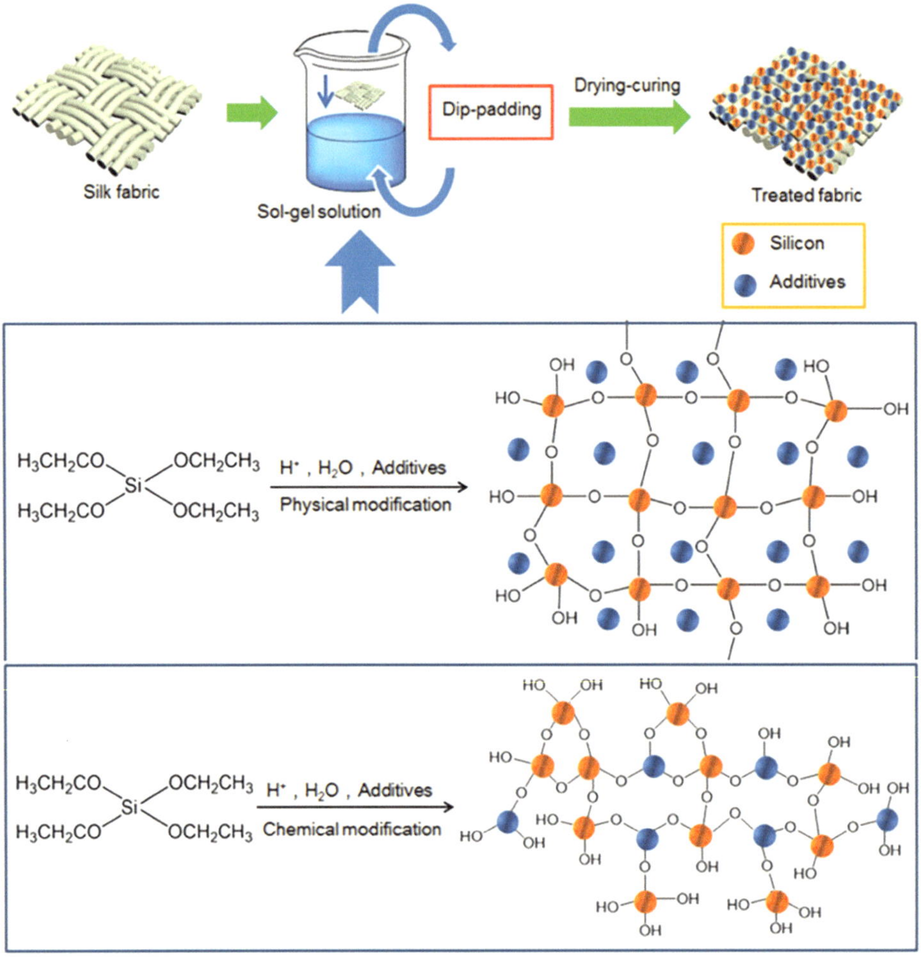

Figure 4.
Approach to fabric functionalization via sol-gel dip-padding process. Reproduced from MDPI ref. [44].

of the fabric such as flexibility, breathability and strength. The use of water-resistant finishes is intended for everyday items like tablecloths, pricey silk garments, uniforms, protective apparel, filter fleece and carpet. Utilizing a straightforward water-based sol-gel procedure, MohdZa'im et al. successfully created a hydrophobic coating for polyester fabric. As a precursor, hexyltrimethoxysilane (HTMS) was diluted in a solution of water, ethanol, and sulfuric acid. The coated polyester fabric's water contact angle was measured at 136.2 degrees, and the HTMS sol-gel coating achieved the development of excellent surface morphology with undetectable fractures that can induce surface roughness on the coated polyester fabric. As a result, these qualities contribute to gaining a textile with hydrophobic properties without altering the fabric's inherent qualities, particularly its softness, breathability and smoothness [51].

Fluoroalkyl silanes were used to further increase the surface water repellency. A different method was developed by Simončič et al. using commercially available aqueous organic–inorganic hybrid precursors as finishing agents, such as fluoroalkyl-functional siloxane (FAS) and 3-(trimethoxysilyl)-propyldimethyloctadecyl ammonium chloride (SiQAC). Two application processes, a one-step treatment and a two-step treatment, were used to apply the hybrid sols to the cotton fabric using the dip-coating and pad-dry-cure methods. One-step treatments (coating FAS-SiQAC) employed a sol mixture of both precursors, whereas two-step treatments (coating SiQAC + FAS) used SiQAC sol and FAS sol. However, in cotton materials, both treatments demonstrated superhydrophobic, oleophobic and antibacterial characteristics [52].

The ECHA committees most recently proposed to restrict the use of particular perfluoroalkyl compounds (PFAS) in certain application sectors. As a result, examples of environmentally friendly, fluorine-free textile finishings with stain- and water-repellent properties are also documented.

In order to create oleophobic textiles with effective stain resistance, Lei et al. discussed the alteration of fabric substrates with an organic silicone tris-trimethylsilypropyl (M_3T) containing methacrylate copolymer [4]. Additionally, Yu et al. used the dry-pad-cure method to cross-link an amino long-chain alkyl polysiloxane functional agent to create a water-repellent fluorine-free cotton fabric [53].

Modified silica hydrosols were employed by Xu et al. to create superhydrophobic cotton fabric. In order to develop a homogenous emulsion, sodium dodecylbenzenesulfonate (SDBS) and methyltrimethoxy silane (MTMS) were dissolved in 100 mL of water and mixed. The solution was then supplemented with hexadecyltrimethoxysilane (HDTMS) while being constantly agitated. Using the dip-pad-cure method, the modified SiO_2 hydrosols were employed to coat some cotton fabric specimens using a single step. The SiO_2 nanoparticles' surface roughness and the HDTMS modification's low surface energy worked together to give the coated cotton fabrics outstanding superhydrophobicity [54].

Functional alkyl(trialkoxy)silane-modified hybrid nanostructured coatings for cotton fabrics were successfully produced by Sfameni et al., specifically via the sol-gel process and pad/dry/cure applications, featuring hydrophobic and water-based stain resistance (**Figure 5**).

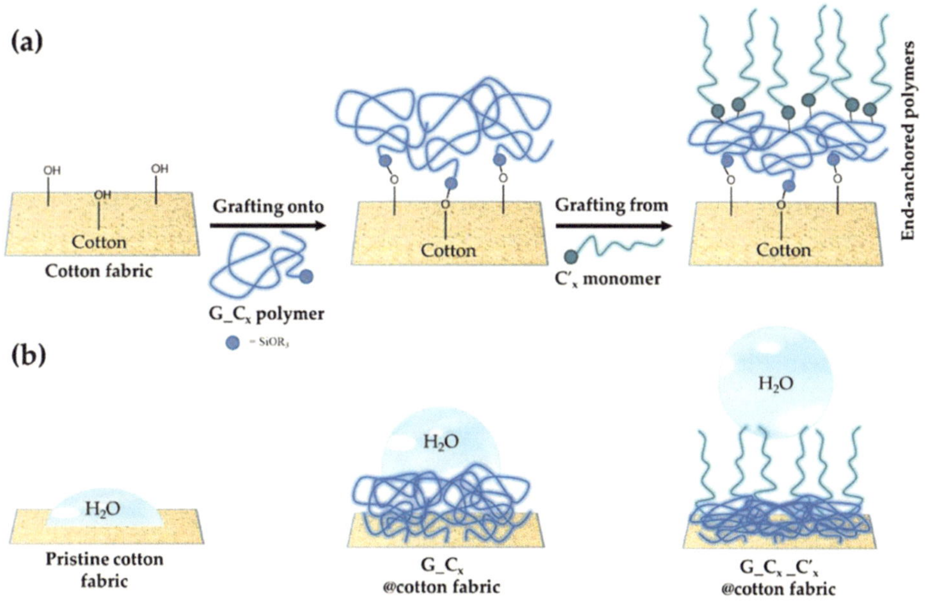

Figure 5.
Cotton fabric functionalization with alkyl(trialkoxy)silane polymer shell by "grafting to" or "grafting from" covalent grafting techniques (a) for the obtaining of hydrophobic and water-based stain-resistant coatings (b). Reproduced from MDPI ref. [55].

The specific objective of the study was to investigate different functional alkyl(trialkoxy)silanes as precursors to synthesize efficient and stable hybrid sol-gel (3-glycidyloxypropyl)trimethoxysilane (GPTMS) based coatings and further reduce cotton surface energy to enhance textile hydrophobicity and water-based stain resistance. In fact, this entire synthetic process has been demonstrated to be a simple, affordable and environmentally friendly method, making it suitable for potentially beneficial application for finishing and functionalizing common textiles in the future [55].

Flame-retardancy is another feature that can be exploited by the design of proper sol-gel based coatings [56]. Textile end uses are restricted and constrained in a wide range of applications because of their possible inherent flammability tendency. For cotton fabric finishing using the sol-gel process, numerous flame-retardant agents have been utilized including halogen-containing (chloride, bromide), nitrogen-containing (melamine, urea), boron and phosphorus compounds [57]. The development of more eco-friendly formaldehyde-free flame-retardant coatings is therefore necessary to reduce the environmental impact of these treatments.

Castellano et al. developed a novel coating for cotton fabrics based on phosphorus, nitrogen and silicon compounds that exhibit self-extinguishing features. In particular, the formaldehyde- and halogen-free coating was synthetized by the use of (3-glycidyloxypropyl)triethoxysilane modified with N-(phosphonomethyl) iminodiacetic acid (PGPTES) and co-hydrolysed and co-condensated with tetraethylorthosilicate (TEOS) as silane cross-linker [58].

Sol-gel process can be exploited also using different inorganic precursors like titanium. Titanium (IV) butoxide, in ethanol, hydrochloric acid and water solution, was employed by Bentis et al. for the formulation of a functional sol with boric acid to obtain sol-based boron-doped titania. Key qualities of cotton samples, in particular their appearance and crystallinity, are unaffected after the application of the TiO_2-boron-based coating by the pad-dry-cure process. The treated specimens demonstrated the ability to absorb more heat than the raw cloth because of the creation of a protective layer that serves as a physical barrier. However, the vertical fire test showed that the coated textiles could demonstrate their fire resistance in actual environments due to their self-extinguishing properties [59].

Drug release and wearable sensors are a recent trend in smart textile research [60–63]. A thermoresponsive polymer, Pluronic F-127, a non-ionic triblock copolymer of poly(ethylene oxide)-poly(propylene oxide)-poly(ethylene oxide) and pH-responsive polymers N,N,N-trimethyl chitosan and polyethylene glycosylated hyaluronic acid, were combined by Chatterjee et al. to create a dual-responsive hydrogel (pH/temperature). Gallic acid was added to the obtained functional hydrogel to aid at it use in textile-based transdermal therapy as a potential medication for the treatment of atopic dermatitis [64].

In regard to other smart applications, a healthcare monitoring system using wearable temperature sensors based on lanthanum-doped aluminum-oxide dielectrics operating at low voltage and high frequency was developed by Park et al. by the sol-gel process on polyimide substrate [65]. Textile substrates can be also coated with such functional sol-gel based formulations to obtain active and stimuli-responsive surfaces for wearable sensors as well i.e. with the use of organic dye chromogenic materials [66]. Cotton fabric coatings with electrical conductivity were moreover created by Trovato et al. using silica sol-gel precursors that had been doped with vertically aligned carbon nanotubes nanofillers dispersed in the functional sol with the help of surfactants in an organic solvent-free aqueous solution [67].

2.2 Blue-growth

Metal products in seawater must fight with two main processes: corrosion and fouling. Corrosion is a serious issue that might seriously compromise the integrity of the product. Meanwhile, biofouling can affect the ship weight and increase fuel costs by the colonization of organisms (bacteria, algae, plants, and other organisms) on wet surfaces. The two phenomena are also complementary to one another [68].

For the period of the building's necessary life, any structure must be protected to withstand corrosive forces to avoid damage caused by the corrosion process. Every method that can slow down or inhibit anodic or cathodic processes, or that can close the conductive channel that connects anodic and cathodic sites, can be utilized to reduce, or stop corrosion. The surface protection process typically entails two (pre) treatments that are applied in succession: the preparation pretreatment, which cleans the metal to be protected from any impurities or remnants of earlier corrosive operations and the actual protection treatment [69].

Despite their high efficacy, phosphating and chromating are costly processes with a significant environmental impact. This is because they use toxic materials like phosphates and chromium, which react with metal supports and form toxic sludge, which must be disposed of in large quantities [70]. It also involves using a lot of thermal energy to heat the reaction baths to the proper temperature. There are numerous additional realistic approaches to lower the rate of corrosion for metallic constructions. Utilizing nobler materials, such as stainless-steel rather than mild carbon steel, can help to naturally form a protective layer (passive layer) on the surface. However, due to cost and processing issues, using noble materials is not always the best option. In reality, applying coatings is an additional method of corrosion prevention. In particular, corrosion caused by exposure to a more or less harsh environment over time can be prevented or delayed by protective coatings which operate as an artificial protective layer on the metal surface. Protective coatings can be made by employing a variety of materials and polymers [71].

To replace potentially harmful chromate-based pretreatment layers, silane-based coatings and films coated with organic and inorganic corrosion inhibitors are now frequently used in corrosion mitigation surface treatment. In this regard, different examples of innovative and sustainable coatings for metal surface protection are reported in the literature. Balestriere et al. developed a sol-gel based formulation based on TEOS and MTMS incorporating borosilicate bioactive glass in order to enhance the anticorrosive and surface performance of stainless-steel implants by a double-layer system obtained through a dip-coating approach [72].

Different additives and nanofillers can be also incorporated in sol-gel based formulations to improve their protective and anticorrosive features like carbon-based nanomaterials and natural derivatives. Different examples are reported in literature about the use for example of graphene oxide (GO) as functional nanofiller. Graphene oxide (GO) is a carbon-based nanomaterial characterized by different oxygen functional groups (such -OH and -COOH) on its basal planes and edges and has received a lot of attention in coating applications as its functional groups ensure its strong compatibility with matrices and additives. Additionally, GO has a great barrier property because of its high aspect-ratio flake-like structure [73]. The effects of GO-filled sol-gel sealing on the corrosion resistance and paint adhesion of anodized aluminum were investigated by Ye et al.. In particular, the GO-filled sol was obtained by a mixture of organosiloxane sol and zirconium alkoxides sol. Anhydrous ethanol,

GPTMS and GO aqueous solution were mixed for the preparation of the organosiloxane sol, meanwhile the Zr sol was prepared by mixing anhydrous ethanol, ethyl acetoacetate and tetrapropyl zirconate. The coating was then applied on aluminum substrates by a dip-dry-cure approach, revealing good anticorrosion performances and paint adhesion [74].

Flavonoids and coumarin derivatives are examples of polyphenolic molecules that exhibit a wide range of features as well as biological and chemical activity. They are typically found as glycosides in nature, particularly in plants, to ensure their water solubility and facilitate their absorption by the organisms. They may perform a variety of functional functions, such as antibacterial and antimicrobial properties, sun protection, or antioxidant activity. The use of such natural compounds with good ligand characteristics is a hot research issue in the area of metal corrosion inhibition [75].

Tamarind shell tannin-doped hybrid sol-gel coatings based on GPTMS and TEOS were developed by Abdulmajid et al. for the improvement of the corrosion protection of mild steel in the acidic medium [76].

In the area of corrosion resistance, phytic acid is another natural derivative that has a wide range of uses because of its non-toxicity. The alloy's corrosion resistance is increased by the chemical adsorption of phytic acid which forms a dense network structure on the alloy's surface and prevents the anion from coming into contact with the substrate [77]. A functional sol based on GPTMS and (3-aminopropyl)triethoxysilane (APTES) was employed in combination with GO and phytic acid for the production of anticorrosive coatings for steel and aluminum substrates [78].

Regarding marine biofouling, which is the other significant issue affecting materials in the marine environment, new coatings and strategies for fouling mitigation have been developed during the past several years [79].

The two most developed types of antifouling coatings include biocide-release coatings and non-biocide-release coatings (**Figure 6**), also indicated as antifouling (AF) and fouling-release (FR) coatings.

In particular, the biocide-release coatings are based on the dispersion of biocides from various polymeric hosting matrices and they release biocides into saltwater gradually over time.

The non-biocidal strategy, related to the fouling release, however, can act in one of two ways:

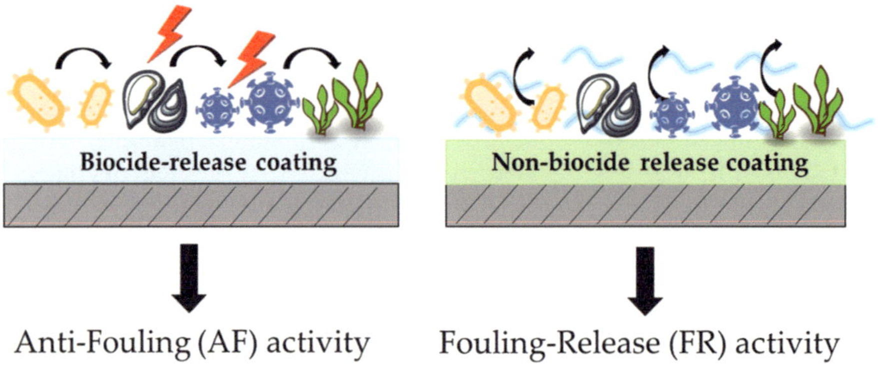

Figure 6.
Types of antifouling coatings for the marine environment. Reproduced from MDPI ref. [61].

i. the separation of stabilized biofoulants, which aims to lessen the force with which microorganisms stick to surfaces in order to facilitate their removal through the weight of the deposits or the flow of water produced during navigation;

ii. the prevention of the attachment of biofoulants, which attempts to prevent the development of a stable fouling coating and, in turn, the adhesion of organic molecules that would start the bio-settlement process.

Moreover, sol-gel coating technology is probably one of the most pertinent tools for the development of environmentally friendly antifouling and fouling release formulations given the current social expectation that new clean, flexible and effective solutions will be adopted rather than the previously mentioned harmful ones.

Numerous techniques have been developed to give innovatively designed coatings to directly interfere with microbe adherence because of topography or surface chemistry, demonstrating fouling release activity, due to the limitations of utilizing biocides in antifouling marine systems. By mixing silicones with fluoropolymers, the critical surface tension can be reduced. Additionally, extensive research has been done on developing hydrophobic surfaces that can prevent the onset of microfouling settlements using proper sol-gel AF-FR formulations and employing different alkyl alkoxysilanes, that is hexadecyltrimethoxysilane, triethoxy(octyl)silane and triethoxy(ethyl)silane (**Figure 7**) [80] and either fluorinated alkoxysilanes [61].

Two novel organoalkoxysilanes were also synthetized by Tan et al. including 2-(2-hydroxy-3-(3-(trimethoxysilyl)propoxy)propyl)benzo[d]isothiazol-3(2H)-one and (N-methoxyacylethyl)-3-aminopropyltriethoxysilane. A series of AF coatings featuring zwitterionic and antibacterial (1,2-benzisothiazolin-3-one) functionalities were obtained through the combination of the two mentioned precursors and TEOS [81]. Given the possible environmental implications of nanomaterial-based nanofillers, surfaces with microstructures incorporating nanomaterials have been widely employed for marine antifouling coatings. Natural cellulose-derived nanofillers have been also explored for antifouling coating preparation. In this regard, Duan et al. developed a bioinspired superhydrophobic coating based on cellulose nanocrystals, CNCs, which are naturally occurring nanomaterials widely employed for

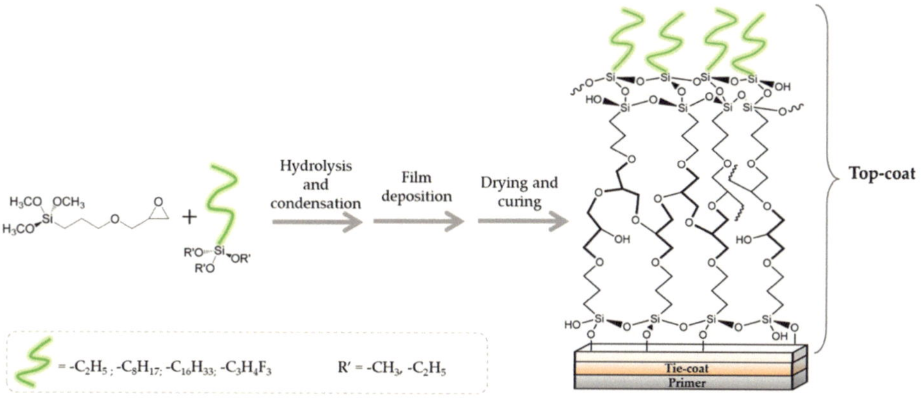

Figure 7.
Hydrophobic and foul-release sol-gel based coating was developed using different alkyl alkoxysilanes. Reproduced from MDPI ref. [80].

the improvement of the mechanical features of materials by their incorporation in polymeric blends [82, 83]. In detail, a sol-gel method with the precursors tetrapropyl zirconate, GPTMS and MTMS, was employed for facile synthesis of an eco-friendly marine antifouling coating incorporating CNCs applied with the spin coating approach in glass substrates [84].

Sol-gel functional coatings can feature also both anticorrosion and antifouling properties [85]. Zhang et al. produced a sol-gel based silver nanoparticle/polytetrafluorethylene coating for stainless-steel with enhanced antibacterial and anticorrosive properties employing titanium (IV) butoxide as precursor and silver nanoparticles [86]. In another study, Zhang et al. (developed a low-surface-free-energy coating based on GO and a fluorine-silicon copolymer featuring self-healing properties for anticorrosion and antifouling applications [87].

2.3 Cultural heritages and buildings

Finally, the protection of architectural and cultural heritage artifacts (based on stone, wood, glass, metals and textiles) should be further mentioned, in order to provide a comprehensive view of the application of novel and environmentally friendly sol-gel based coatings for various surfaces, beyond those made of textiles and metals. Over the last several decades, a variety of protective coatings have been developed to maintain the integrity of cultural heritages and stop the degrading process by reducing the pace of deterioration brought on by weathering and environmental processes (**Figure 8**).

The ideal coating for such an application sector must feature different qualities like transparency, reversibility, compatibility with the surface, long lifetime, ease of synthesis, low-cost maintenance and non-toxicity [89].

Considering these features, sol-gel based coatings find a place as the best recent innovative research results, featuring protective and multifunctional performances, as well as cultural heritages. Due to its advantageous characteristics, including its high chemical stability, non-toxicity, strong photoreactivity and affordability, ZnO is widely used in a variety of technical fields. Due to its antibacterial activity, it is recently

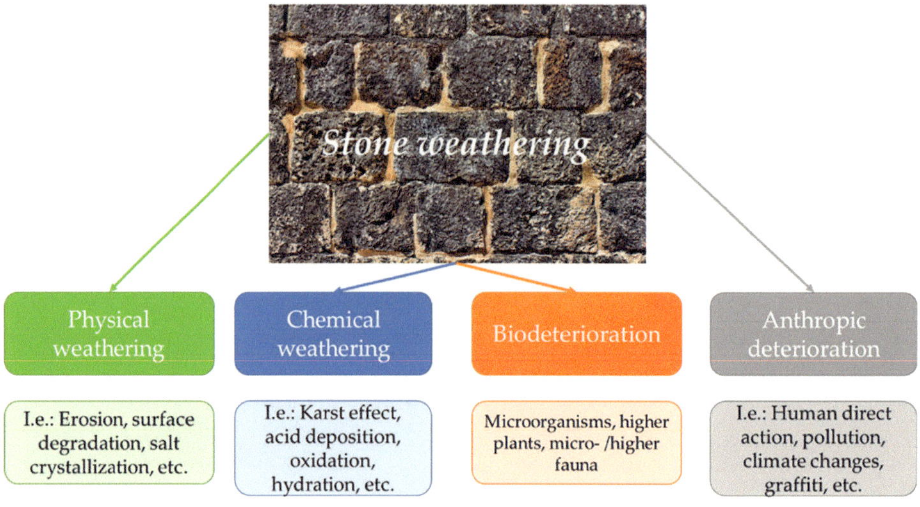

Figure 8.
Weathering phenomena that may cause the stone deterioration. Reproduced from MDPI ref. [88].

useful in preventing the biodeterioration of cultural heritage structures. Doped ZnO nanostructures provide superior antibacterial capabilities than their undoped analogues as reported in some examples. Weththimuni et al. developed a polydimethylsiloxane (PDMS)/ZrO_2-doped ZnO nanocomposites as a protective and self-cleaning coating for stone materials applied by the brushing method. ZrO_2-ZnO core-shell nanoparticles were synthetized and incorporated with octamethylcyclotetrasiloxane and CsOH, employed as a catalyst for the ring-opening polymerization and subsequently, hexamethyldisiloxane to produce in-situ the ZrO_2-ZnO-PDMS nanocomposite [90]. In another recent work, an isobutyltriethoxysilane modified silica (IBTES@SiO_2) nano-coating was produced by Wang et al. to protect from water, ultraviolet radiation and the destructive effect of condensation some outdoor sandstone cultural relics. In this regard, the functional coating featuring self-cleaning, porous and superhydrophobic properties was obtained with a one-step spraying method starting from a formulation based on isobutyltriethoxysilane and nano-SiO_2 prepared at room temperature [91]. A multifunctional protective formulation was developed by Azadi et al. containing methyl methacrylate, 3-(trimethoxysilyl)propyl methacrylate, perfluorooctyltrichlorosilane and TiO_2 to improve the weathering resistance, self-cleaning properties and hydrophobic behavior of the coated stones with the final nanohybrid film [92].

Glass cultural heritages represent another class of interest in the protective coating development field. Some methods have already been employed to protect glass surfaces from mechanical and climatic hazards such as passive glass object protection with an exterior glazing system placed in front of the stained-glass window. Other techniques involve directly applying chemical inhibitors and consolidants to the surface of the glass after appropriate treatment. Generally, a good protective coating should satisfy some conservation needs (maintenance of the original aspect of the objects and their optical properties) and enhance certain properties (resistance to aggressive environments, slowing down the rate of degradation processes) [93].

Historical glazed wall tiles are an uncommon kind of outdoor artistic expression that is used in structures from many different countries; hence they are frequently prone to biodeterioration. The study carried out by Coutinho et al. was determined whether protective coatings for glazed tiles might be used to stop biological colonization. The anti-biofouling qualities of thin coatings of titanium dioxide produced by sol-gel using titanium (IV) isopropoxide as the precursor in ethanol and acidic water (HNO_3), were tested after their application by spin coating on glazed tiles in order to determine the sustainability and effectiveness of this approach for use in cultural heritage applications [94].

Also, wood cultural heritage elements must be preserved. It has been recently investigated by Andriulo et al. on how to deacidify and consolidate alum-treated wood using multifunctional organic/inorganic hybrid systems. In detail, the formulation was prepared by mixing poly(dimethylsiloxane) hydroxy-terminated (PDMS-OH), TEOS or MTES. Subsequently, the Nanorestore paper® Propanol 5 and 5 g/L $Ca(OH)_2$ nanoparticles dispersion in 2-propanol was added to the functional sol. The designed formulations, which combine nanostructured $Ca(OH)_2$ with a network of polysiloxanes, have shown promise in terms of penetration, deacidification, consolidation and preservation of wood artefact characteristics [95].

3. Conclusions

Nowadays, the advancements of nanotechnology in the field of innovative material and implemented surfaces must face the more sustainable expectations and environmental needs of modern society. Despite the new rules and laws about the limitation of the use of toxic and harmful additives for surface finishings, today there is still the challenge to find solutions that are safe for the environment and human health because also of the large use of organic solvents.

In this regard, sol-gel technology aims to achieve environmentally friendly, easy and innovative solutions to address these tasks, by employing different approaches mainly related to water-based or alcohol-based formulations. This chapter described the most recent developments in the field of inorganic–organic hybrid materials and the importance of the choice of different inorganic precursors and functional additives to fine-tune the final desired product properties, also by means of sol-gel synthesis. In particular, the use of titanium, zirconium, zinc and silane-based formulations is reported and moreover, their functionalization with different functional alkoxysilanes, nanofillers (i.e. carbon-based nanomaterials) or functional natural molecules. Thus, their applications for the implementation of the protective, hydrophobic, anticorrosive, flame-retardant, drug release, antifouling and antibacterial features of different surfaces (including metals, textiles, wood, glass and stones) are shown. Therefore, the described solutions may address important advancements in functional and smart textiles, cultural heritage protection and blue-growth applications.

Acknowledgements

CNR and MUR are gratefully acknowledged for the financial support.

Conflict of interest

The authors declare no conflict of interest.

Author details

Silvia Sfameni[1], Giulia Rando[1,2*] and Maria Rosaria Plutino[1*]

1 Department of ChiBioFarAm, University of Messina, Institute for the Study of Nanostructured Materials, ISMN—CNR, Palermo, Messina, Italy

2 Department of Chemical, Biological, Pharmaceutical and Environmental Science (ChiBioFarAm), University of Messina, Messina, Italy

*Address all correspondence to: giulia.rando@unime.it and mariarosaria.plutino@cnr.it

References

[1] Zheng S, Chen H, Tong X, Wang Z, Crittenden JC, Huang M. Integration of a photo-Fenton reaction and a membrane filtration using CS/PAN@FeOOH/g-C3N4Electrospun nanofibers: Synthesis, characterization, self-cleaning performance and mechanism. Applied Catalysis B: Environmental. 2021;**281**:119519. Available from: https://www.sciencedirect.com/science/article/pii/S0926337320309346

[2] Wang C, Makvandi P, Zare EN, Tay FR, Niu L. Advances in antimicrobial organic and inorganic nanocompounds in biomedicine. Advances in Therapy. 2020;**2020**:3

[3] Rosu C, Lin H, Jiang L, Breedveld V, Hess DW. Sustainable and long-time 'rejuvenation' of biomimetic water-repellent silica coating on polyester fabrics induced by rough mechanical abrasion. Journal of Colloid and Interface Science. 2018;**516**:202-214 Available from: https://www.sciencedirect.com/science/article/pii/S0021979718300638

[4] Lei H, Xiong M, Xiao J, Zheng L, Zhu Y, Li X, et al. Fluorine-free low surface energy organic coating for anti-stain applications. Progress in Organic Coatings. 2017;**103**:182-192 Available from: https://www.sciencedirect.com/science/article/pii/S0300944016305264

[5] Chen G, Yuan B, Wang Y, Chen X, Huang C, Shang S, et al. Nacre-biomimetic graphene oxide paper intercalated by phytic acid and its ultrafast fire-alarm application. Journal of Colloid and Interface Science. 2020;**578**:412-421 Available from: https://www.sciencedirect.com/science/article/pii/S0021979720307281

[6] Wang W, Meng Q, Li Q, Liu J, Zhou M, Jin Z, et al. Chitosan derivatives and their application in biomedicine. International Journal of Molecular Sciences. 2020;**2020**:21

[7] Lim JYC, Goh SS, Liow SS, Xue K, Loh XJ. Molecular gel sorbent materials for environmental remediation and wastewater treatment. Journal of Materials Chemistry A. 2019;**7**:18759-18791. DOI: 10.1039/C9TA05782J

[8] Wright JD, Sommerdijk NAJM. Sol-Gel Materials: Chemistry and Applications. Boca Raton, FL, USA: CRC Press; 2018

[9] Guo Y, Xu K, Wu C, Zhao J, Xie Y. Surface chemical-modification for engineering the intrinsic physical properties of inorganic two-dimensional nanomaterials. Chemical Society Reviews. 2015;**44**:637-646. DOI: 10.1039/C4CS00302K

[10] Sikri A, Sikri J. Nanomaterials: A boon in prosthodontics. EC Dent. Sci. 2022;**21**:1-2

[11] Sanchez C, Belleville P, Popall M, Nicole L. Applications of advanced hybrid organic–inorganic nanomaterials: From laboratory to market. Chemical Society Reviews. 2011;**40**:696-753. DOI: 10.1039/C0CS00136H

[12] Pandey S, Mishra SB. Sol–gel derived organic–inorganic hybrid materials: Synthesis, characterizations and applications. Journal of Sol-Gel Science and Technology. 2011;**59**:73-94. DOI: 10.1007/s10971-011-2465-0

[13] García-Martínez JM, Collar EP. Organic–inorganic hybrid materials. Polymer. 2021;**2021**:13

[14] Rando G, Sfameni S, Galletta M, Drommi D, Cappello S, Plutino MR.

Functional Nanohybrids and nanocomposites development for the removal of environmental pollutants and bioremediation. Molecules. 2022;**2022**:27

[15] Schubert U. Chemistry and fundamentals of the sol–gel process. Sol-Gel Handbook. 2015;**2015**:1-28. DOI: 10.1002/9783527670819.ch01

[16] Cavallaro G, Micciulla S, Chiappisi L, Lazzara G. Chitosan-based smart hybrid materials: A physico-chemical perspective. Journal of Materials Chemistry B. 2021;**9**:594-611. DOI: 10.1039/D0TB01865A

[17] Mazari SA, Ali E, Abro R, Khan FSA, Ahmed I, Ahmed M, et al. Nanomaterials: Applications, waste-handling, environmental toxicities, and future challenges – A review. Journal of Environmental Chemical Engineering. 2021;**9**:105028. Available from: https://www.sciencedirect.com/science/article/pii/S2213343721000063

[18] Jeon IY, Baek JB. Nanocomposites derived from polymers and inorganic nanoparticles. Materials. 2010;**3**:3654-3674

[19] Al Zoubi W, Kamil MP, Fatimah S, Nashrah N, Ko YG. Recent advances in hybrid organic-inorganic materials with spatial architecture for state-of-the-art applications. Progress in Materials Science. 2020;**112**:100663. Available from: https://www.sciencedirect.com/science/article/pii/S007964252030027X

[20] Sfameni S, Hadhri M, Rando G, Drommi D, Rosace G, Trovato V, et al. Inorganic finishing for textile fabrics: Recent advances in Wear-resistant, UV protection and antimicrobial treatments. Inorganics. 2023;**2023**:11

[21] Zanurin A, Johari NA, Alias J, Mas Ayu H, Redzuan N, Izman S. Research progress of sol-gel ceramic coating: A review. Materials Today Proceedings. 2022;**48**:1849-1854 Available from: https://www.sciencedirect.com/science/article/pii/S2214785321060053

[22] Huang G, Huo L, Jin Y, Yuan S, Zhao R, Zhao J, et al. Fluorine-free superhydrophobic PET fabric with high oil flux for oil–water separation. Progress in Organic Coatings. 2022;**163**:106671. Available from: https://www.sciencedirect.com/science/article/pii/S0300944021005427

[23] Rosace G, Cardiano P, Urzì C, De Leo F, Galletta M, Ielo I, et al. Potential roles of fluorine-containing sol-gel coatings against adhesion to control microbial biofilm. IOP Conference Series Materials Science Engineering. 2018;**459**:12021. DOI: 10.1088/1757-899X/459/1/012021

[24] Ielo I, Giacobello F, Castellano A, Sfameni S, Rando G, Plutino MR. Development of antibacterial and antifouling innovative and eco-sustainable sol–gel based materials: From marine areas protection to healthcare applications. Gels. 2022;**2022**:8

[25] Catauro M, Vecchio CS. In: Demetzos C, Pippa N, editors. Sol-Gel Synthesis and Characterization of Hybrid Materials for Biomedical Applications BT - Thermodynamics and Biophysics of Biomedical Nanosystems: Applications and Practical Considerations. Singapore: Springer Singapore; 2019. pp. 445-475. DOI: 10.1007/978-981-13-0989-2_13

[26] Tao Y, Pescarmona PP. Nanostructured oxides synthesised via scCO2-assisted sol-gel methods and their application in catalysis. Catalysts. 2018;**2018**:8

[27] Rao KS, El-Hami K, Kodaki T, Matsushige K, Makino K. A novel method

for synthesis of silica nanoparticles. Journal of Colloid and Interface Science. 2005;**289**:125-131 Available from: https://www.sciencedirect.com/science/article/pii/S0021979705001566

[28] Parashar M, Shukla VK, Singh R. Metal oxides nanoparticles via sol–gel method: A review on synthesis, characterization and applications. Journal of Materials Science: Materials in Electronics. 2020;**31**:3729-3749. DOI: 10.1007/s10854-020-02994-8

[29] Pierre AC. Introduction to Sol-Gel Processing. Cham, Switzerland: Springer Nature Switzerland AG; 2020

[30] Soloviev A, Jensen H, Søgaard EG, Kanaev AV. Aggregation kinetics of sol-gel process based on titanium tetraisopropoxide. Journal of Materials Science. 2003;**38**:3315-3318. DOI: 10.1023/A:1025198323886

[31] Elanany M, Selvam P, Yokosuka T, Takami S, Kubo M, Imamura A, et al. A quantum molecular dynamics simulation study of the initial hydrolysis step in sol–gel process. The Journal of Physical Chemistry. B. 2003;**107**:1518-1524. DOI: 10.1021/jp026816z

[32] Butt MA. Thin-film coating methods: A successful marriage of high-quality and cost-effectiveness—a brief exploration. Coatings. 2022;**2022**:12

[33] Beig B, Liaqat U, Niazi MF, Douna I, Zahoor M, Niazi MB. Current challenges and innovative developments in hydroxyapatite-based coatings on metallic materials for bone implantation: A review. Coatings. 2020;**2020**:10

[34] Yan Y, Li J, Liu Q, Zhou P. Evaporation effect on thickness distribution for spin-coated films on rectangular and circular substrates. Coatings. 2021;**2021**:11

[35] Lu S, Shao J, Wu F. Industrial applications of sol–gel derived coatings. Journal of Sol-Gel Science Technology. 2022;**2022**. DOI: 10.1007/s10971-022-05988-6

[36] Baino F, Fiume E, Miola M, Verné E. Bioactive sol-gel glasses: Processing, properties, and applications. International Journal of Applied Ceramic Technology. 2018;**15**:841-860. DOI: 10.1111/ijac.12873

[37] Uhlmann DR, Teowee G. Sol-gel science and technology: Current state and future prospects. Journal of Sol-Gel Science and Technology. 1998;**13**:153-162. DOI: 10.1023/A:1008692430779

[38] Nisticò R, Scalarone D, Magnacca G. Sol-gel chemistry, templating and spin-coating deposition: A combined approach to control in a simple way the porosity of inorganic thin films/coatings. Microporous and Mesoporous Materials. 2017;**248**:18-29 Available from: https://www.sciencedirect.com/science/article/pii/S1387181117302573

[39] Júnior HL, Neves RM, Monticeli FM, Dall AL. Smart fabric textiles: Recent advances and challenges. Text. 2022;**2**:582-605

[40] Meng X, Zhang K, Guo X, Wang C, Sun L. Preparation of micro-textures on cemented carbide substrate surface by plasma-assisted laser machining to enhance the PVD tool coatings adhesion. Journal of Materials Processing Technology. 2021;**288**:116870. Available from: https://www.sciencedirect.com/science/article/pii/S0924013620302843

[41] Periyasamy AP, Venkataraman M, Kremenakova D, Militky J, Zhou Y. Progress in sol-gel Technology for the Coatings of fabrics. Materials (Basel). 2020;**2020**:13

[42] Trovato V, Sfameni S, Rando G, Rosace G, Libertino S, Ferri A, et al. A review of stimuli-responsive smart materials for wearable Technology in Healthcare: Retrospective, perspective, and prospective. Molecules. 2022;**2022**:27

[43] Bakar NHA, Yusop HM, Ismail WN, Zulkifli NF. Sol-Gel Finishing for Protective Fabrics. Biointerface Research in Applied Chemistry. 2022;**13**:283-299. DOI: 10.33263/BRIAC133.283

[44] Liu C, Xing T, Wei B, Chen G. Synergistic effects and mechanism of modified silica sol flame retardant systems on silk fabric. Materials (Basel). 2018;**2018**:11

[45] Liu S, Wan C, Chen Y, Chen R, Zhang F, Zhang G. A novel high-molecular-weight flame retardant for cotton fabrics. Cellulose. 2020;**27**:350-615. DOI: 10.1007/s10570-020-03020-0

[46] Nguyen-Tri P, Tran HN, Plamondon CO, Tuduri L, Vo DVN, Nanda S, et al. Recent progress in the preparation, properties and applications of superhydrophobic nano-based coatings and surfaces: A review. Progress in Organic Coatings. 2019;**132**:235-256 Available from: https://www.sciencedirect.com/science/article/pii/S030094401930092X

[47] Kumar A, Tudu BK, Pandit SK. Development of novel anti-wetting coating on cellulosic surface using low carbon butyric acid. Cellulose. 2021;**28**:4824-4834. DOI: 10.1007/s10570-021-03797-8

[48] Chauhan P, Kumar A. Development of a microbial coating for cellulosic surface using aloe vera and silane. Carbohydr. Polym. Technol. Appl. 2020;**1**:100015. Available from: https://www.sciencedirect.com/science/article/pii/S2666893920300153

[49] Paras KA. Smart bioinspired anti-wetted surfaces: Perspectives, fabrication, stability and applications. Current Research Green Sustainable Chemistry. 2021;**4**:100139. Available from: https://www.sciencedirect.com/science/article/pii/S2666086521000862

[50] Chouirfa H, Bouloussa H, Migonney V, Falentin-Daudré C. Review of titanium surface modification techniques and coatings for antibacterial applications. Acta Biomaterialia. 2019;**83**:37-54 Available from: https://www.sciencedirect.com/science/article/pii/S1742706118306354

[51] Zaim NN, Yusop HM, Ismail WN. Synthesis of water-repellent coating for polyester fabric. Emerging Science Journal. 2021;**5**:747-754

[52] Simončič B, Tomšič B, Černe L, Orel B, Jerman I, Kovač J, et al. Multifunctional water and oil repellent and antimicrobial properties of finished cotton: Influence of sol–gel finishing procedure. Journal of Sol-Gel Science and Technology. 2012;**61**:350-354. DOI: 10.1007/s10971-011-2633-2

[53] Yu C, Shi K, Ning J, Zheng Z, Yu H, Yang Z, et al. Preparation and application of fluorine-free finishing agent with excellent water Repellency for cotton fabric. Polymers. 2021;**2021**:13

[54] Xu L, Zhuang W, Xu B, Cai Z. Superhydrophobic cotton fabrics prepared by one-step water-based sol–gel coating. Journal of the Textile Institute. 2012;**103**:311-319. DOI: 10.1080/00405000.2011.569238

[55] Sfameni S, Lawnick T, Rando G, Visco A, Textor T, Plutino MR. Functional Silane-based Nanohybrid materials for the development of hydrophobic and water-based stain resistant cotton fabrics coatings. Nanomaterials. 2022;**2022**:12

[56] Malucelli G. Sol-gel and layer-by-layer coatings for flame-retardant cotton fabrics: Recent advances. Coatings. 2020;**10**:333

[57] Mazumder NUS, Islam MT. Flame retardant finish for textile fibers. In: Innovative and Emerging Technologies for Textile Dyeing and Finishing. Beverly, MA, USA: Scrivener Publishing; 2021. pp. 373-405. DOI: 10.1002/9781119710288.ch13

[58] Castellano A, Colleoni C, Iacono G, Mezzi A, Plutino MR, Malucelli G, et al. Synthesis and characterization of a phosphorous/nitrogen based sol-gel coating as a novel halogen- and formaldehyde-free flame retardant finishing for cotton fabric. Polymerase Degradation Stable. 2019;**162**:148-159. DOI: 10.1016/j.polymdegradstab.2019.02.006

[59] Bentis A, Boukhriss A, Gmouh S. Flame-retardant and water-repellent coating on cotton fabric by titania–boron sol–gel method. Journal of Sol-Gel Science and Technology. 2020;**94**:719-730. DOI: 10.1007/s10971-020-05224-z

[60] Chatterjee S, Hui PC. Review of applications and future prospects of stimuli-responsive hydrogel based on Thermo-responsive biopolymers in drug delivery systems. Polymers. 2021;**2021**:13

[61] Sfameni S, Rando G, Galletta M, Ielo I, Brucale M, De Leo F, et al. Design and development of fluorinated and biocide-free sol–gel based hybrid functional coatings for anti-biofouling/foul-release activity. Gels. 2022;**2022**:8

[62] Puoci F, Saturnino C, Trovato V, Iacopetta D, Piperopoulos E, Triolo C, et al. Sol-gel treatment of textiles for the entrapping of an antioxidant/anti-inflammatory molecule: Functional coating morphological characterization and drug release evaluation. Applied Sciences. 2020;**2020**:10

[63] Libertino S, Plutino MR, Rosace G. Design and development of wearable sensing nanomaterials for smart textiles. AIP Conference Proceedings. 2018;**2018**:1990

[64] Chatterjee S, Hui PC, leung, Kan C wai, Wang W. Dual-responsive (pH/temperature) Pluronic F-127 hydrogel drug delivery system for textile-based transdermal therapy. Scientific Reports. 2019;**9**:11658. DOI: 10.1038/s41598-019-48254-6

[65] Park SJ, Jeon JY, Kang BC, Ha TJ. Wearable temperature sensors based on lanthanum-doped aluminum-oxide dielectrics operating at low-voltage and high-frequency for healthcare monitoring systems. Ceramics International. 2021;**47**:4579-4586 Available from: https://www.sciencedirect.com/science/article/pii/S0272884220330625

[66] Trovato V, Mezzi A, Brucale M, Rosace G, Plutino MR. Alizarin-functionalized organic-inorganic silane coatings for the development of wearable textile sensors. Journal of Colloid and Interface Science. 2022;**617**:463-477 Available from: https://www.sciencedirect.com/science/article/pii/S002197972200385X

[67] Trovato V, Teblum E, Kostikov Y, Pedrana A, Re V, Nessim GD, et al. Electrically conductive cotton fabric coatings developed by silica sol-gel precursors doped with surfactant-aided dispersion of vertically aligned carbon nanotubes fillers in organic solvent-free aqueous solution. Journal of Colloid and Interface Science. 2021;**586**:120-134

[68] Cerchier P, Pezzato L, Gennari C, Moschin E, Moro I, Dabalà M. PEO

coating containing copper: A promising anticorrosive and antifouling coating for seawater application of AA 7075. Surface and Coatings Technology. 2020;**393**:125774. Available from: https://www.sciencedirect.com/science/article/pii/S0257897220304436

[69] Bouali AC, Serdechnova M, Blawert C, Tedim J, Ferreira MGS, Zheludkevich ML. Layered double hydroxides (LDHs) as functional materials for the corrosion protection of aluminum alloys: A review. Applied Materials Today. 2020;**21**:100857. Available from: https://www.sciencedirect.com/science/article/pii/S235294072030305X

[70] Doerre M, Hibbitts L, Patrick G, Akafuah NK. Advances in automotive conversion coatings during Pretreatment of the body structure: A review. Coatings. 2018;**2018**:8

[71] Verma A, Jain N, Rastogi S, Dogra V, Sanjay SM, Siengchin S, et al. Mechanism, anti-corrosion protection and components of anti-corrosion polymer coatings. In: Polymer Coatings. Boca Raton, FL, USA: CRC Press; 2020. pp. 53-66

[72] Balestriere MA, Schuhladen K, Herrera Seitz K, Boccaccini AR, Cere SM, Ballarre J. Sol-gel coatings incorporating borosilicate bioactive glass enhance anti corrosive and surface performance of stainless steel implants. Journal of Electroanalytical Chemistry. 2020;**876**:114735. Available from: https://www.sciencedirect.com/science/article/pii/S1572665720309632

[73] Gao Y, Fan Y, Zhang J, Liu X, Wang N, Yang S. The study of graphene oxide on the regulations and controls of the sol-gel film structure and its performance. Meta. 2022;**2022**:12

[74] Yu M, Dong H, Shi H, Xiong L, He C, Liu J, et al. Effects of graphene oxide-filled sol-gel sealing on the corrosion resistance and paint adhesion of anodized aluminum. Applied Surface Science. 2019;**479**:105-113 Available from: https://www.sciencedirect.com/science/article/pii/S016943321930337X

[75] Veys-Renaux D, Reguer S, Bellot-Gurlet L, Mirambet F, Rocca E. Conversion of steel by polyphenolic model molecules: Corrosion inhibition mechanism by rutin, esculin, esculetol. Corrosion Science. 2018;**136**:1-8 Available from: https://www.sciencedirect.com/science/article/pii/S0010938X17300872

[76] Abdulmajid A, Hamidon TS, Hussin MH. Tamarind shell tannin-doped hybrid sol–gel coatings on mild steel in acidic medium toward improved corrosion protection. J. Coatings Technol. Res. 2022;**19**:527-542. DOI: 10.1007/s11998-021-00539-0

[77] Zhang M, Chen R, Liu X, Liu Q, Liu J, Yu J, et al. Anticorrosion study of phytic acid ligand binding with exceptional self-sealing functionality. Journal of Alloys and Compounds. 2020;**818**:152875. Available from: https://www.sciencedirect.com/science/article/pii/S0925838819341210

[78] Sfameni S, Del Tedesco A, Rando G, Truant F, Visco A, Plutino MR. Waterborne eco-sustainable sol–gel coatings based on Phytic acid intercalated graphene oxide for corrosion protection of metallic surfaces. International Journal of Molecular Sciences. 2022;**2022**:23

[79] Callow JA, Callow ME. Trends in the development of environmentally friendly fouling-resistant marine coatings. Nature Communications. 2011;**2**:244. DOI: 10.1038/ncomms1251

[80] Sfameni S, Rando G, Marchetta A, Scolaro C, Cappello S, Urzì C, et al.

Development of eco-friendly hydrophobic and fouling-release coatings for blue-growth environmental applications: Synthesis, mechanical characterization and biological activity. Gels. 2022;**2022**:8

[81] Tan J, Liang X, Yang J, Zhou S. Sol–gel-derived hard coatings from tetraethoxysilane and organoalkoxysilanes bearing zwitterionic and isothiazolinone groups and their antifouling behaviors. Journal of Materials Chemistry B. 2022;**10**:406-417. DOI: 10.1039/D1TB02069B

[82] Kassab Z, Aziz F, Hannache H, Ben Youcef H, El Achaby M. Improved mechanical properties of k-carrageenan-based nanocomposite films reinforced with cellulose nanocrystals. International Journal of Biological Macromolecules. 2019;**123**:1248-1256 Available from: https://www.sciencedirect.com/science/article/pii/S0141813018354485

[83] Cao L, Liu C, Zou D, Zhang S, Chen Y. Using cellulose nanocrystals as sustainable additive to enhance mechanical and shape memory properties of PLA/ENR thermoplastic vulcanizates. Carbohydrate Polymers. 2020;**230**:115618. Available from: https://www.sciencedirect.com/science/article/pii/S014486171931286X

[84] Duan Y, Wu J, Qi W, Su R. Eco-friendly marine antifouling coating consisting of cellulose nanocrystals with bioinspired micromorphology. Carbohydrate Polymers. 2023;**304**:120504. Available from: https://www.sciencedirect.com/science/article/pii/S0144861722014096

[85] Jin H, Wang J, Tian L, Gao M, Zhao J, Ren L. Recent advances in emerging integrated antifouling and anticorrosion coatings. Materials and Design. 2022;**213**:110307. Available from: https://www.sciencedirect.com/science/article/pii/S0264127521008625

[86] Zhang S, Liang X, Gadd GM, Zhao Q. A sol–gel based silver nanoparticle/polytetrafluorethylene (AgNP/PTFE) coating with enhanced antibacterial and anti-corrosive properties. Applied Surface Science. 2021;**535**:147675. Available from: https://www.sciencedirect.com/science/article/pii/S0169433220324326

[87] Zhang H, Liang T, Liu Y, Misra RDK, Zhao Y. Low-surface-free-energy GO/FSiAC coating with self-healing function for anticorrosion and antifouling applications. Surface and Coatings Technology. 2021;**425**:127690. Available from: https://www.sciencedirect.com/science/article/pii/S0257897221008641

[88] Fistos T, Fierascu I, Doni M, Chican IE, Fierascu RC. A short overview of recent developments in the application of polymeric materials for the conservation of stone cultural heritage elements. Materials (Basel). 2022;**2022**:15

[89] Artesani A, Di Turo F, Zucchelli M, Traviglia A. Recent advances in protective coatings for cultural heritage–an overview. Coatings. 2020;**2020**:10

[90] Weththimuni ML, Ben CM, Tredici I, Licchelli M. Polydimethylsiloxane (PDMS)/ZrO2-doped ZnO nanocomposites as protective coatings for stone materials. In: International Conference on Metrology for Archaeology and Cultural Heritage. Trento, Italy; 2020

[91] Wang G, Chai Y, Li Y, Luo H, Zhang B, Zhu J. Sandstone protection by using nanocomposite coating of silica. Applied Surface Science. 2022:156193

Available from: https://www.sciencedirect.com/science/article/pii/S0169433222037217

[92] Azadi N, Parsimehr H, Ershad-Langroudi A. Cultural heritage protection via hybrid nanocomposite coating. Plastics, Rubber and Composites. 2020;**49**:414-424. DOI: 10.1080/14658011.2020.1784589

[93] De Bardi M, Hutter H, Schreiner M, Bertoncello R. Sol–gel silica coating for potash–lime–silica stained glass: Applicability and protective effect. Journal of Non-Crystalline Solids. 2014;**390**:45-50 Available from: https://www.sciencedirect.com/science/article/pii/S0022309314001057

[94] Coutinho ML, Veiga JP, Macedo MF, Miller AZ. Testing the feasibility of titanium dioxide sol-gel coatings on Portuguese glazed tiles to prevent biological colonization. Coatings. 2020;**2020**:10

[95] Andriulo F, Vespignani L, Steindal CC, Bortolini M, de Ferri L. Evaluation of sol-gel hybrid nanocomposites for dry medieval wood. Journal of Cultural Heritage. 2022;**56**:96-107 Available from: https://www.sciencedirect.com/science/article/pii/S129620742200098X

Chapter 7

Sol-Gel Production of Semiconductor Metal Oxides for Gas Sensor Applications

Walid Belaid, Amina Houimi, Shrouk E. Zaki and Mohamed A. Basyooni

Abstract

As they are widely utilized in industries including the food packaging industry, indoor air quality testing, and real-time monitoring of man-made harmful gas emissions to successfully combat global warming, reliable and affordable gas sensors represent enormous market potential. For environmental monitoring, chemical safety regulation, and many industrial applications, the detection of carbon monoxide (CO), carbon dioxide (CO_2), nitrogen dioxide (NO_2), and methane (CH_4) gases is essential. To reliably and quantitatively detect these gases, much-improved materials and methods that are adaptable to various environmental factors are needed using low-cost fabrication techniques such as sol-gel. The advantages of employing metal oxide nanomaterials-based chemoresistive for creating high-performance gas sensors are shown by key metrics such as selectivity, sensitivity, reaction time, and detection. The primary sensing methods are also grouped and thoroughly covered. In light of the current constraints, anticipated future developments in the field of sol-gel nanomaterial-based chemoresistive gas sensors are also highlighted.

Keywords: gas sensor, sol-gel, nanowires, semiconductor metal oxides, thin film

1. Introduction

The sol-gel method has been a topic of interest since its inception as a chemical method for creating glasses at lower temperatures than traditional melting processes. Achieving homogeneity in complex compositions was a requirement that was met by controlling the hydrolysis and condensation reactions of various precursors using different chemical strategies. The method's versatility and precision in controlling material composition, morphology, and properties have made it highly attractive for various applications, such as catalysts, sensors, optics, and energy conversion devices. A wide variety of materials, including inorganic membranes, monolithic glasses and ceramics, thin films, ultrafine powders, and hybrid materials, are frequently produced using the sol-gel process. The hydrolysis of a precursor solution yields suspended colloidal particles as the foundation of the sol-gel technique. Subsequent condensation of the particles leads to the formation of a gel-like substance, which can be further processed to produce the desired material [1–3]. In recent years, the sol-gel technique has emerged as a promising method for depositing gas sensors. This technique offers several advantages over other deposition methods, including low-temperature processing, high purity of the deposited material, and the ability to control the thickness and porosity of the film. Sol-gel-based gas sensors have been used in a wide range of applications, including environmental monitoring, automotive exhaust detection, and medical diagnostics [4–12].

Gas sensors have emerged as an indispensable technology in various industries due to their ability to detect and measure different gases in different environments. For instance, in the petrochemical industry, the detection of hazardous gases such as CO [13, 14], CO_2 [15–19], NO_2 [20, 21], and CH_4 [22, 23] is critical to maintaining safe working conditions and avoiding catastrophic accidents. Similarly, food processing industries require gas sensors to monitor for hazardous gases, such as carbon dioxide, that can accumulate in confined spaces and pose a serious threat to worker safety. Environmental monitoring also relies on gas sensors to detect pollutants and greenhouse gases, like CO_2 and CH_4, that contribute to climate change [24]. In medical applications, gas sensors are used for monitoring gas concentrations in breath analysis and anesthesia delivery. Every year, approximately 3.8 million people suffer from serious and potentially deadly illnesses caused by air pollution. Additionally, household pollution is responsible for roughly 20% of cardiovascular and stroke-related fatalities [25]. Therefore, the importance of gas sensors in different applications cannot be overstated, as they play a significant role in protecting human health and safety [26–32].

The purpose of this chapter is to offer a broad overview of gas sensors prepared using the sol-gel technique for the detection of CO, CO_2, NO_2, and CH_4 gases based on metal oxides. The chapter will provide an introduction to the fundamental principles of gas sensing, the properties of sol-gel films, and the mechanisms of gas detection. Recent advancements in sol-gel technology for the deposition of gas sensors, including the use of various precursors, dopants, and modifiers, will also be discussed. The references cited in this chapter represent a selective but informative collection of papers that describe the fundamental principles of synthesis choices. The goal of the chapter is to offer readers an essential and comprehensive overview of gas-sensing applications using the sol-gel method.

2. Gas sensor classification

Gas sensors can be broadly categorized based on their gas-sensing materials and response. The primary types include optical, electrochemical, and electrical sensors, as demonstrated in an overview of gas sensor classification depicted in **Figure 1**. Optical gas sensors use light to detect the presence of gases by measuring changes in light absorption or scattering [33, 34]. In contrast, electrochemical gas sensors use a chemical reaction to produce an electrical signal that can be measured and interpreted as the concentration of gas [35]. On the other hand, electrical gas sensors use a change in electrical resistance to indicate the presence of a gas by reacting with it and altering its electrical properties [36].

In addition to the gas sensor types mentioned above, a range of additional sensor technologies exists for gas detection purposes, as summarized in **Table 1** which includes information on the principle of detection and typical gases detected for various gas sensor technologies.

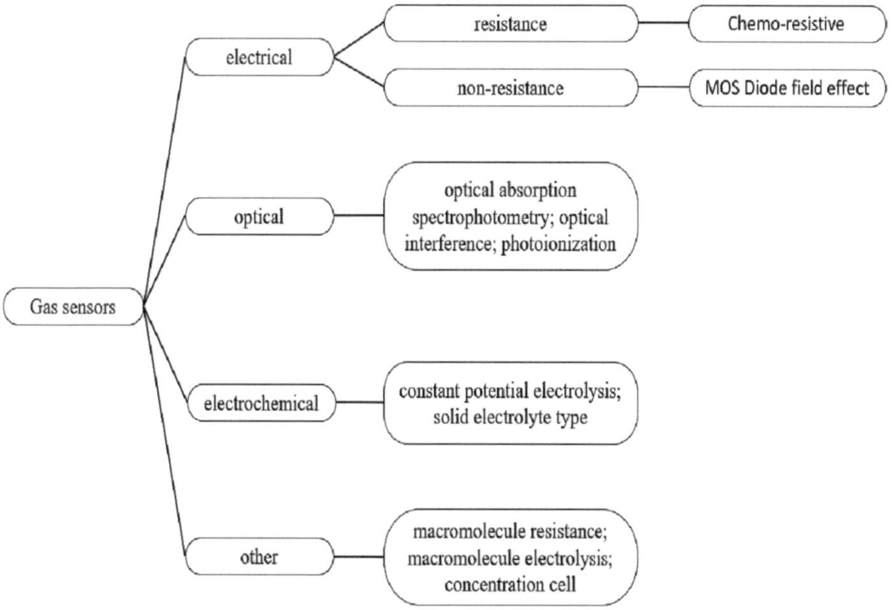

Figure 1.
Overview of gas sensor classification.

Gas sensor technology	Principle of detection	Typical gases detected
Optical	Light absorption or scattering	Carbon monoxide, carbon dioxide, nitrogen dioxide, ozone, sulfur dioxide, volatile organic compounds
Electrochemical	Chemical reaction producing an electrical signal	Carbon monoxide, nitrogen dioxide, hydrogen sulfide, sulfur dioxide, chlorine, ozone
Electrical	Change in electrical resistance	Carbon monoxide, methane, propane, hydrogen, natural gas
Conductometric	Change in electrical conductivity	Flammable and explosive gases
Thermal Conductivity	Change in thermal conductivity	Hydrogen, helium
Piezoelectric	Change in mechanical stress	Volatile organic compounds, other gases
Surface Acoustic Wave	Change in mass or conductivity	Toxic and explosive gases
Photoionization	Ionization of gas molecules by ultraviolet light	Volatile organic compounds, other gases
Solid-state	Change in electrical conductivity or resistance	Carbon monoxide, nitrogen dioxide, methane

Table 1.
Classification of gas sensor technologies based on the principle of detection, with examples of typical gases detected.

Currently, there is an extensive research focused on the development of electrical nanosensors with enhanced performance [37–39]. One type of electrical nanosensor is the resistive gas sensor [40], which is an electronic device that alters its electrical resistance according to the surrounding gas ambient. Resistive gas sensors are categorized into two types based on their operating principle: chemical and physical [41]. Physical resistive gas sensors are capable of detecting changes in the physical properties of the gas, such as temperature, pressure, or mass, while chemoresistive gas sensors [42] employ a sensing layer to interact with the target gas. Although the fabrication and optimization of resistive gas sensors remain challenging due to the complexity of the materials involved and the interactions between the gas and sensing layer, chemoresistive sensors are widely used for gas detection due to their high sensitivity and selectivity. They offer several advantages, such as low cost, low power consumption, and miniaturization capability, making them suitable for various portable and wearable applications. Chemoresistive sensors are also highly sensitive and selective toward specific gases, enabling them to detect even low concentrations of gases [43]. In the next section, we will delve deeper into the mechanisms and applications of chemoresistive gas sensors, which will shed more light on their importance in the field of gas sensing.

3. Chemoresistive gas sensors: mechanisms, advantages, and applications

Chemoresistive gas sensors utilize changes in the electrical conductivity or resistance of the sensing material to detect various chemical or gaseous analytes. Typically, the sensing material is composed of a semiconductor made from a metal oxide, which may be classified as either p-type or n-type.

In n-type gas-sensing materials, the presence of gas molecules leads to the formation of surface states, which capture electrons from the conduction band and reduce the free carrier concentration, increasing resistance. The magnitude of this change in resistance depends on several factors, including the concentration and type of gas, the temperature, and the properties of the sensing material. In p-type gas-sensing materials, the presence of gas molecules leads to the release of holes from the valence band, which increases the free carrier concentration and result in a decrease in resistance [44]. As with n-type sensing materials, the magnitude of this change in resistance depends on several factors, including the concentration and type of gas, the temperature, and the properties of the sensing material [45].

Both n and p types of gas-sensing materials can be used in chemoresistive gas sensors to detect a wide variety of gases and chemical analytes. The performance of these sensors can be affected by factors such as the method of sensing material deposition, the type of sensing material used, and the operating temperature. However, ongoing research in this area continues to improve the sensitivity, selectivity, and other performance characteristics of chemoresistive gas sensors [46]. Additionally, creating a junction between p-type and n-type oxide semiconductors using various contact arrangements can offer innovative approaches to developing gas sensors as reported in an excellent review on nanoscale metal oxide-based heterojunctions for gas sensing by Miller et al. [47]. Chemoresistive gas sensors offer several advantages, such as simplicity, low cost, and potential for miniaturization. Compared to other gas sensors, they do not require reference electrodes, which can complicate the sensor design and increase the production cost. Additionally, chemoresistive gas sensors offer fast response times and can detect gases over a wide concentration range [38, 48]. However, chemoresistive gas sensors may have some limitations compared to other types of sensors. For instance, they may not offer the same level of selectivity as electrochemical sensors, which use a chemical reaction to detect specific gases.

Additionally, chemoresistive gas sensors may not be able to detect certain types of gases that are detectable using optical sensors, which use light to detect changes in the concentration of gases. Despite these limitations, chemoresistive gas sensors are widely used in a range of applications, including medical diagnosis [49], environmental monitoring [50], and industrial safety [51]. Ongoing research continues to improve the sensing performance and capabilities of chemoresistive gas sensors, making them a promising option for gas detection in various fields.

4. Gas sensing operating temperatures

Temperature is a crucial aspect of gas sensing and has a significant impact on the performance of gas sensors. The sensitivity and selectivity of gas sensors vary depending on the temperature range in which they operate. Operating temperatures that are high can heighten the sensitivity and reaction rates of gas sensors. However, they may lead to thermal drift and instability. On the other hand, low operating temperatures can improve sensor stability but decrease sensitivity and response time. Due to their larger bandgap, metal oxide sensors typically require high operating temperatures between 100 and 450°C for surface redox reactions and reaction kinetics to facilitate sensing measurements [52, 53]. Nevertheless, this high-temperature operation has limitations in terms of energy conservation, application, and potential hazards such as gas explosions and sensor instability. Therefore, operating gas sensors at room temperature is highly desirable to minimize the risk of a gas explosion, decrease energy consumption, increase sensor life, and enable integration into smartphone devices [25].

The two dimensional transition metal dichalcogenides [54, 55] and low-dimensional structures [56] generally operate at lower temperatures compared to metal oxide sensors, typically in the range of 25–200°C. This lower-temperature operation can offer advantages such as reduced power consumption, improved stability, and wider application in portable devices. However, it can also lead to lower sensitivity and slower response times compared to sensors operating at higher temperatures. Therefore, understanding and optimizing the operating temperature of gas sensors is crucial for achieving accurate and reliable gas detection. The gas-sensing performance of sol-gel deposited gas sensors is highly dependent on the operating temperature, which affects the sensor's sensitivity, selectivity, response time, recovery time, and ability to detect specific gases. **Table 2** summarizes the advantages and disadvantages of different gas-sensing operating temperatures.

5. Metal oxides nanomaterials for gas-sensing applications

The development of gas-sensing materials with high sensitivity and selectivity has been a significant research focus in recent years. Among the various synthesis methods, the sol-gel technique has gained important attention as a promising approach for depositing different types of materials onto various substrates. A schematic view of the sol-gel process for the development of ZnO-based films is given in **Figure 2**. Sol-gel processing offers several advantages, such as low-temperature processing, precise control over the deposited material's composition and morphology, and scalability. By tailoring the deposition parameters and optimizing the material properties, sol-gel-deposited gas-sensing materials have shown superior sensing performance compared to conventionally synthesized materials.

Temperature	Advantages	Disadvantages
Low	• Increases efficiency and lifespan of some electronic and mechanical components • Improves stability and accuracy of some sensors • Reduces risk of thermal runaway and fire hazards in some applications	• Increases viscosity or freezing of some fluids • Reduces battery capacity • Increases susceptibility to condensation and moisture • Some materials become brittle or lose their elasticity
Room Temperature	• Most convenient and cost-effective option • Most components are designed and tested for room-temperature operation • Many sensors and instruments optimized for performance and stability at this range • Comfortable and safe for human operators	• Reduced sensitivity or selectivity of some sensors • Increased noise or drift in some electronic components • Reduced efficiency or performance in some energy conversion systems
High	• Increases efficiency and power density of some energy conversion systems • Improves performance and sensitivity of some sensors • Reduces risk of contamination or corrosion in some applications	• Increases thermal stress and mechanical wear on some components • Reduces the lifetime and stability of some sensors and electronic devices • Increases risk of thermal runaway or fire hazards in some systems • Requires specialized materials and thermal management, which increases the cost and complexity of the system

Table 2.
Advantages and disadvantages of different gas-sensing operating temperatures.

Metal oxide nanomaterials deposited by the sol-gel method have been used for the detection of a wide range of gases, including toxic gases, flammable gases, and environmental pollutants. These sensors have several advantages, including high sensitivity, fast response time, low-power consumption, and the ability to operate at room temperature [57, 58]. Metal oxide-based gas sensors have been extensively studied and are currently the most investigated type of gas sensor. Recently, there has been a growing trend in using these materials with sizes ranging from 1 to 100 nm for gas sensing due to their size-dependent properties [59, 60]. These nanomaterials possess unique mechanical, optical, electrical, catalytic, and magnetic properties, and have high surface area per unit mass with new emerging physical and chemical properties. As the size of the material decreases, the specific surface area and surface-to-volume ratio increase significantly. Additionally, the size and geometry of the semiconductor nanomaterials can affect the movement of electrons and holes. The greenhouse gases sensors are largely studied using metal oxide materials.

5.1 Carbon dioxide

Gas sensors can be broadly categorized based on their gas-sensing materials and response. The primary types include optical, electrochemical, and electrical

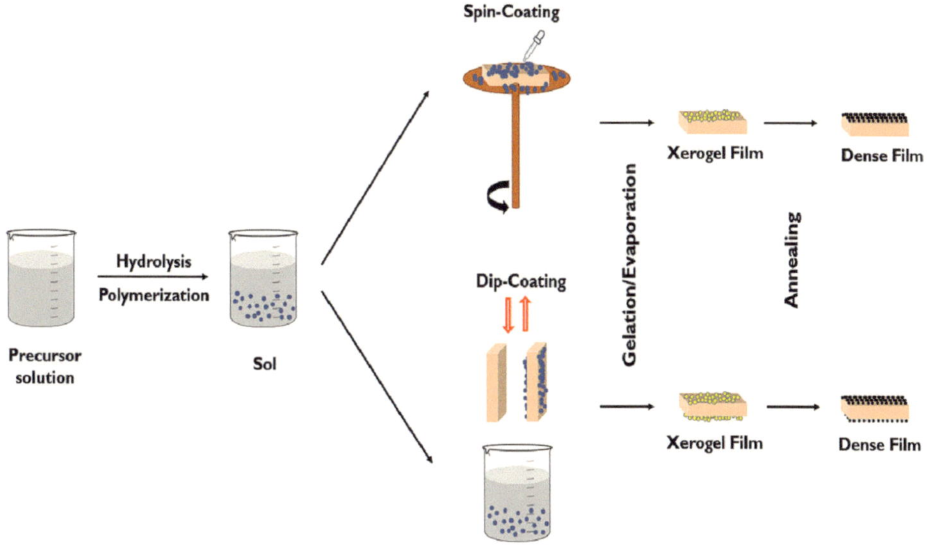

Figure 2.
Schematic of sol-gel process for ZnO thin film deposition.

sensors, as demonstrated in **Table 1**. CO_2 emissions contribute significantly to global warming and climate change, making it crucial to detect and regulate its emission. Metal oxide-based CO_2 sensors can be cost-effective and efficient to measure CO_2 levels even at room temperature (**Table 3**). Nanostructured nickel ferrite ($NiFe_2O_4$) thin films were tested as liquefied CO_2 gas sensors at a low temperature of 27°C [66]. The thin films, deposited using the spin-coating technique, had 1.3 sensitivity at 1000 ppm CO_2 levels with 100 and 400 s as response and recovery time, respectively. Panday et al. [61] tested another metal oxide for CO_2 detection. Sb-doped SnO_2 nanostructured thin films were deposited using the same technique with a 450°C annealing temperature. At room temperature (30°C), the nondoped SnO_2 thin film exhibits an optimum gas response of 78.5%.

Material	Shape	Concentration (ppm)	Response (R)/ Sensitivity(S)	Response/ Recover time	Temperature (°C)	Ref.
SnO_2:Sb	Nanostructured thin films	—	1.78 (R)	2.6/5.8 s	30	[61]
ZnO/Na	Nanostructured thin film	50	81.9 (R)	283/472 s	RT	[62]
TiO_2:PANI	Thin film	1000	53 (R)	9.2/5.7 min	30	[63]
Y@ ZnO:CdO	Thin film	500	9 (R)	4/2 s	RT	[64]
SnO_2:Co_3O_4	Nanocomposite	—	13.68 (R)	2/12 s	30	[18]
$PrFeO_3$	Nanopowder	1000	8.44 (R)	—	160	[65]
$NiFe_2O_4$	Nanostructured thin films	1000	1.3 (S)	100/400 s	27	[66]

Table 3.
Comparison of metal oxides based-CO_2 gas sensing performance.

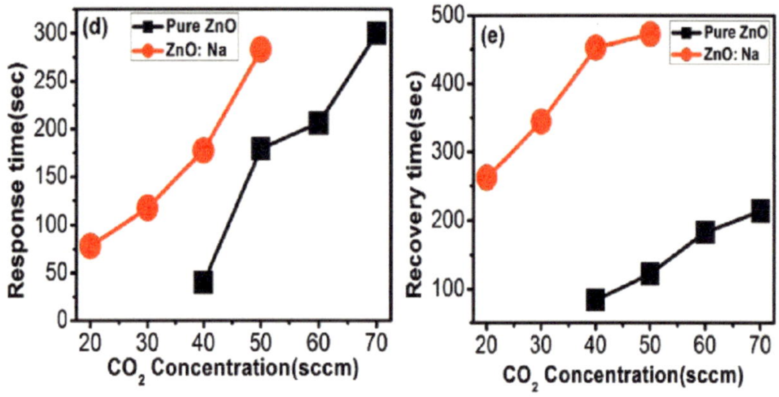

Figure 3.
Response/recovery time of ZnO and Na-ZnO-based CO_2 gas sensors [62].

Y-doped-ZnO: CdO nanocomposite thin films prepared by sol-gel spin coating technique also proved its efficiency for CO_2 detection [64]. With their cauliflower-like morphology, these metal oxides showed a sensor response of nine at room temperature. A mixture of two metal oxides was also used for the detection of CO_2 gas in the work presented by Joshi et al. [18]. The nanocomposite porous mixture of SnO_2-Co_3O_4 (1:2) thin films showed a high response of 13.68 at a temperature of 30°C compared to the other used ratios. Response/recovery time is an important parameter indicating how fast the sensor can detect the gas, for instance, toward CO_2 as it is the most important greenhouse gas **Figure 3** for ZnO [62].

5.2 Carbon monoxide

CO gas detection is crucial in protecting against the potentially deadly effects of carbon monoxide poisoning, which can be caused by the incomplete combustion of fossil fuels. Early detection can save lives and prevent serious health consequences, making CO gas detection a vital component of any safety plan. Metal oxides synthesized using the sol-gel technique were largely investigated for CO-sensing properties (see **Table 4**).

Material	Shape	Concentration (ppm)	Response (R)/ Sensitivity(S)	Response/ Recover time	Temperature	Ref.
SnO_2:Sb	WO_3	50–500	1.01–5.67 (S)	200°C	200°C	[67]
ZnO/Na	In/Pd@SnO_2	1	3 (R)	15/22 s	140°C	[68]
TiO_2:PANI	TiO_2:CeO_2	400	16.1 (R)	32/45 s	200°C	[69]
Y@ ZnO:CdO	TiO_2–ZrO_2	100	9.1 (R)	42/48 s	150°C	[70]
SnO_2:Co_3O_4	Au@ In_2O_3	5	104 (S)	130/50 s	RT	[71]
$PrFeO_3$	TiO_2/ perovskite	400	38.41% (R)	—	200°C	[72]

Table 4.
Comparison of metal oxides based-CO gas sensing performance.

Susanti et al. [67] reported in their work the detection of 50 ppm of CO gas at 200°C by WO_3 nanomaterial. The best results were achieved when the thin film's calcination temperature reached 500°C. In another work, CO gas was detected at room temperature by sol-gel-synthesized WO_3 thin films [73]. The deposition of AuNP was ensured by the dip-coating technique which resulted in high coverage of In_2O_3 nanowires treated with SAM (assembled monolayer) layer. As a result, they witnessed a great enhancement in CO oxidation and an improvement in the sensing capabilities of the device at room temperature [71]. Mixtures of Perovskite oxide ($La_{0.8}Sr_{0.2}Co_{0.5}Ni_{0.5}O_3$) with other metal oxides (ZnO: Al [74], ZnO [75], TiO_2 [72], SnO_2 [76]) were reported as well for the detection of carbon monoxide at relatively low temperatures that ranges between 180 and 200°C.

5.3 Nitrogen dioxide

NO_2 gas detection is crucial for identifying and mitigating the harmful effects of nitrogen dioxide on human health and the environment. It allows for prompt action to be taken to reduce emissions, protect public health, and promote clean air. Overall, NO_2 gas detection is vital for safeguarding well-being and ensuring a healthy environment. We present in **Table 5** a performance comparison of NO_2 gas sensors.

Due to their unique effective surface area, WO_3 nanostructures such as nanoparticles have been examined as excellent candidates for gas sensors at lower temperatures. In a study done by Yan et al. [77], WO_3 nanoparticles were synthesized using a sol-gel method onto porous silicon and alumina substrates. Gas sensing tests showed that the WO_3 nanoparticles/porous silicon exhibited improved NO_2-sensing properties at room temperature compared to WO_3 on alumina. In another study, using a sol-gel method graphene-wrapped WO_3 nanosphere composite was synthesized [83]. This composite exhibited p-type gas-sensing behavior, with a linear response to NO_2 concentration at room temperature. Tin dioxide showed a good response toward NO_2 gas in different works [81, 82]. Recently, Kumar et al. [82] reported a remarkably high sensitivity of SnO_2 metal oxide at room temperature toward 2 ppm of NO_2 gas.

Material	Shape	Concentration (ppm)	Response (R)/ Sensitivity (S)	Response/ Recover time	Temperature	Ref.
WO_3	Nanoparticle on porous silicon	2	3.27 (R)	2/>20 min	RT	[77]
Ce @NiO	Nanostructured thin film	40	29% (S)	62/595 s	150°C	[78]
Pt@ZnO	Thin film	1	1.02 (S)	420/660 s	200°C	[79]
WO_3: In_2O_3	Mixed nanocomposites	1	~99% (S)	16/– min	~140°C	[80]
In @SnO_2	Nanoparticles	500	72 (R)	—	150°C	[81]
SnO_2	Nanoparticles	2	8.44 (R)	184/432 s	RT	[82]

Table 5.
Comparison of metal oxides based-NO_2 gas-sensing performance.

5.4 Methane

Natural gas's main component, CH_4 gas, is an odorless, colorless, and extremely combustible gas. It frequently serves as a fuel source for appliances that heat and prepare food, as well as for industrial activities. The dangers posed by methane gas make its detection crucial. In small places, it can potentially replace oxygen, causing asphyxiation. Methane gas is also a strong greenhouse gas that, when released into the atmosphere, accelerates climate change.

Metal oxides synthesized with the sol-gel technique played an important role in improving CH_4 gas sensors (**Table 6**). A large number of works investigated the possibility of applying metal oxides synthesized by the sol-gel method for methane gas sensors (**Table 6**); however, only a few of them are operating at low temperatures ($T < 200°C$).

Abruzzi et al. [84] SnO_2 manage to obtain very good performance from SnO_2 nanomaterials-based CH_4 sensors using the sol-gel method. At 80°C operation temperature, the response and recovery times were 29 and 47 s, respectively. ZnO thin films were also tested for methane gas sensors using the sol-gel deposition technique [85]. ZnO nanocrystalline thin films were modified using a Pd atom in the work presented by Bhattacharyya et al. [85]. They reported a good response of ZnO nanocrystalline deposited on SiO_2/n-Si. The Pd-modified ZnO thin films showed lower sensing T of 150°C compared to the unmodified thin films (250°C). Also, it displayed lower response and recovery times.

The production of a nanowrinkled $Zn_{0.92}Fe_{0.08}O$ thin film (see **Figure 4**) utilizing the high rpm electro-spin patterning technology via sol-gel approach is discussed in the work presented by Anchal et al. [86]. The film's nanostructure is homogenous and smooth, making it good for gas adsorption and sensing activities. With a rise in gas concentration and operating temperature, the sensor responds more quickly. The maximum response, which outperforms the ZnO flat thin film-based sensor, is 83.4% for 500 ppm of methane at 200°C. Perovskite oxides based on rare earth metals ($LnFeO_3$) are also being researched for use as methane gas sensors [87].

Material	Shape	Concentration	Response (R)/ Sensitivity (S)	Response/ Recover time	Temperature	Ref.
SnO_2	Nanoparticles	20,000 ppm	61 (%)	29/47 s	80°C	[84]
Pd modified ZnO	Nanocrystalline thin films	1%	87.2%	6/33 s	150°C	[85]
$Zn_{0.92}Fe_{0.08}O$	Nano-wrinkled thin film	500 ppm	83.4%	96/107 s	200°C	[86]
$SmFeO_2$	Nano-crystalline	200 ppm	22% (R)	—	200°C	[87]
SiNWs/TiO2	Core-shell nanostructure	120 ppm	182% (R)	75/191 s	RT	[88]
CoO	Thin films	2000 ppm	—	250/250 s	210°C	[89]

Table 6.
Comparison of metal oxides based-CH_4 gas-sensing performance.

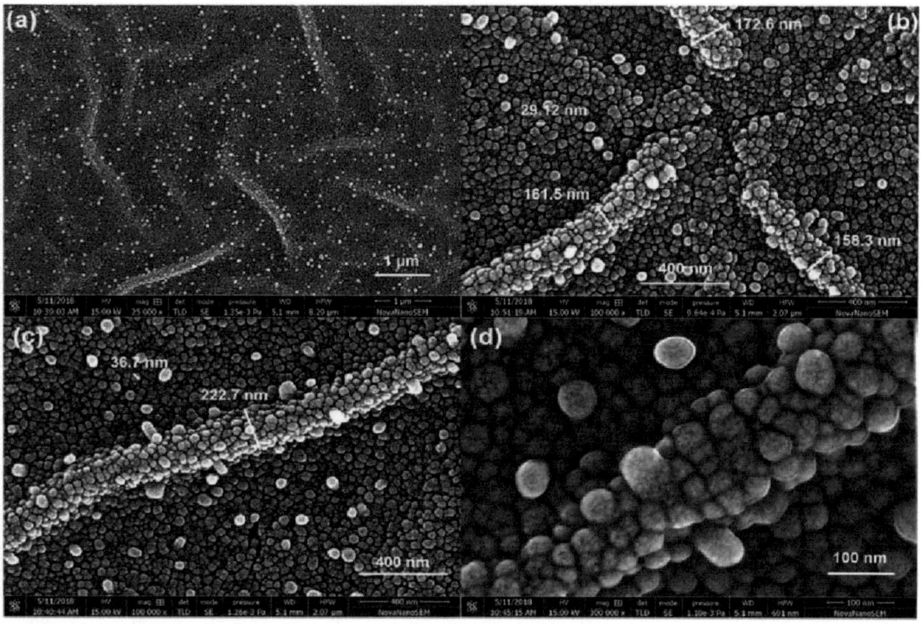

Figure 4.
FE-SEM of nanowrinkled $Zn_{0.92}Fe_{0.08}O$ thin film showing (a) large area view, (b) enlarged structure, (c) grown single wrinkle, (d) wrinkle surface and grain/grain boundary structure [86].

6. Conclusions

The field of sol-gel nanomaterial-based chemoresistive gas sensors especially based on metal oxides has been rapidly evolving in recent years. These sensors have great potential for use in a wide range of applications, including environmental monitoring, medical diagnosis, and industrial process control. One of the current problems in this field is the lack of selectivity of some sensors, which can result in false positives or negatives. It is working to improve the selectivity of these sensors by developing new materials with specific binding sites for target gases. Another challenge is the stability and reproducibility of the sensors, which can be affected by variations in processing conditions and environmental factors such as temperature and humidity. In terms of anticipated future developments, one promising area is the integration of sol-gel nanomaterial-based gas sensors with other technologies, such as microelectronics and wireless communication. This could enable the development of smart sensing systems that can remotely monitor and analyze gas concentrations in real-time. Another area of potential development is the use of sol-gel nanomaterials for the detection of other types of analytes, such as volatile organic compounds and biological molecules. This could open up new applications for these sensors in areas such as food safety, healthcare, and homeland security. Overall, while there are still some challenges to be addressed, the field of sol-gel nanomaterial-based chemoresistive gas sensors holds great promise for the development of highly selective, stable, and versatile sensing systems in the future.

Acknowledgements

All authors equally contributed to this study including the conception and design of the work; the acquisition, analysis, and interpretation of data for the work; participating in drafting or revising the work; and approving the final version of the work to be published.

Conflict of interest

The authors declare no conflict of interest.

Author details

Walid Belaid[1], Amina Houimi[2], Shrouk E. Zaki[3,4] and Mohamed A. Basyooni[2,3,5]*

1 Department of Physics, Institute of Science, University of Selcuk, Konya, Turkey

2 Science and Technology Research and Application Center (BITAM), Necmettin Erbakan University, Konya, Turkey

3 Department of Nanotechnology and Advanced Materials, Graduate School of Applied and Natural Science, Selçuk University, Konya, Turkey

4 Theoretical Physics Department, National Research Center, Dokki, Cairo, Egypt

5 Solar and Space Research Department, National Research Institute of Astronomy and Geophysics (NRIAG), Cairo, Egypt

*Address all correspondence to: m.a.basyooni@gmail.com

References

[1] Guglielmi M, Kickelbick G, Martucci A. Sol-gel nanocomposites. Springer, Elsiver; 2014

[2] Elhadj B, Bahri D, Abdelhalim Z, Wan Jeffrey B, Rabie A, Ammar B, et al. Synthesis and properties of copper doped zinc oxide thin films by sol-gel, spin coating and dipping: A characterization review. Journal of Molecular Structure. 2022;**1267**:133639

[3] Myasoedova TN, Kalusulingam R, Mikhailova TS. Sol-gel materials for electrochemical applications: Recent advances coatings. MDPI Coatings. 2022;**12**(11):1625

[4] Kumar V, Madan R, Singh B, Mohan D. Room-temperature acetone gas sensing properties of graphene oxide/zinc oxide nanocomposites synthesized by sol-gel method. Journal of Materials Science: Materials in Electronics. 2023;**34**(7):582

[5] Xia Z, Zheng C, Hu J, Yuan Q, Zhang C, Zhang J, et al. Synthesis of SnO_2 quantum dot sensitized $LaFeO_3$ for conductometric formic acid gas sensors. Sensors and Actuators B: Chemical. 2023;**379**:133198

[6] Wang S, Xiong H, Tang Y, Zhang W, Zhang Y, Liu Q, et al. High sensitivity and selectivity triethylamine gas sensor based on ZnO—$SmFeO_3$ molecular imprinted polymers. Materials Research Bulletin. May 2023;**161**:112147

[7] Bae M-S, Lee J-W, Koh J-H. Synthesis and characterization of multi-phase structure, optical and electrical properties on (Ga–Sn) oxide composite thin film by sol-gel method. Materials Chemistry and Physics. 2023;**293**:126960

[8] Mahdi HI, Bakr NA, Al-Saadi TM. Preparation and characterization of NixMn0. 25-xMg0. 75Fe_2O_4 Nano-ferrite as NO_2 Gas Sensing Material. arXiv preprint arXiv:230101728. 2023

[9] Tihtih M, Ibrahim J, Basyooni MA, En-Nadir R, Belaid W, Hussainova I, et al. Development of yttrium-doped BaTiO(3) for next-generation multilayer ceramic capacitors. ACS Omega. 2023;**8**(9):8448-8460

[10] Tihtih M, Ibrahim J, Basyooni MA, En-Nadir R, Hussainova I, Kocserha I. Functionality and activity of sol-gel-prepared Co and Fe co-doped Lead-free BTO for Thermo-optical applications. ACS Omega. 2023;**8**(5):5003-5016

[11] Tihtih M, Ibrahim JEFM, Basyooni MA, En-nadir R, Belaid W, Abdelfattah MM, et al. Enhanced optical and thermal conductivity properties of barium titanate ceramic via strontium doping for thermo-optical applications. Optical and Quantum Electronics. 2023;**55**(3):1-20

[12] Tihtih M, Ibrahim JEFM, Basyooni MA, Kurovics E, Belaid W, Hussainova I, et al. Role of A-site (Sr), B-site (Y), and a, B sites (Sr, Y) substitution in lead-free $BaTiO_3$ ceramic compounds: Structural, optical, microstructure, mechanical, and thermal conductivity properties. Ceramics International. 2023;**49**(2):1947-1959

[13] Aranthady C, Shanbhag GV, Sundaram NG. Polyaniline/(Ta(2) O(5)-SnO(2)) hybrid nanocomposite for efficient room temperature CO gas sensing. RSC Advances. 2022;**12**(25):15759-15766

[14] Kim J-H, Mirzaei A, Kim HW, Kim SS. Low-voltage-driven sensors based on ZnO nanowires for

room-temperature detection of NO2 and CO gases. ACS Applied Materials & Interfaces. 2019;**11**(27):24172-24183

[15] Muckley ES, Aytug T, Mayes R, Lupini AR, Carrillo J-MY, Goswami M, et al. Hierarchical TiO_2:Cu_2O nanostructures for gas/vapor sensing and CO_2 sequestration. ACS Applied Materials & Interfaces. 2019;**11**(51):48466-48475

[16] Kalyakin AS, Medvedev DA, Volkov AN. Electrochemical sensors based on proton-conducting electrolytes for determination of concentration and diffusion coefficient of CO_2 in inert gases. Chemical Engineering Science. 2021;**229**:116046

[17] Bouachma S, Ayouz-Chebout K, Kechouane M, Manseri A, Yaddadene C, Menari H, et al. Synthesis of PSi-n/CuO-p/Cu_2O-n heterostructure for CO_2 gas sensing at room temperature. Applied Physics A. 2021;**128**(1):69

[18] Joshi G, Rajput JK, Purohit LP. SnO_2–Co_3O_4 pores composites for CO_2 gas sensing at low operating temperature. Microporous and Mesoporous Materials. 2021;**326**:111343

[19] Lan L, Chen J, Zhao X, Ghasemifard H. VCSEL-based atmospheric trace gas sensor using first harmonic detection. IEEE Sensors Journal. 2019;**19**(13):4923-4931

[20] Wang N, Ye J-X, Sun J-B, Zhang X-F, Deng Z-P, Xu Y-M, et al. Rapid and accurate detection of highly toxic NO_2 gas based on catkins biomass-derived porous In_2O_3 microtubes at low temperature. Sensors and Actuators B: Chemical. 2022;**361**:131692

[21] Wang C, Zhang L, Huang H, Xi R, Jiang DP, Zhang SH, et al. A nanocomposite consisting of ZnO decorated graphene oxide nanoribbons for resistive sensing of NO(2) gas at room temperature. Mikrochimica Acta. 2019;**186**(8):554

[22] Vatandoust L, Habibi A, Naghshara H, Aref SM. Fabrication and investigation of TiO1.5/ZnO nanocomposite nanosensor for detection of CO and CH_4 gases. Surfaces and Interfaces. 2022;**31**:102001

[23] Zhang S, Li Y, Sun G, Zhang B, Wang Y, Cao J, et al. Synthesis of NiO-decorated ZnO porous nanosheets with improved CH4 sensing performance. Applied Surface Science. 2019;**497**:143811

[24] Mar KA, Unger C, Walderdorff L, Butler T. Beyond CO_2 equivalence: The impacts of methane on climate, ecosystems, and health. Environmental Science & Policy. 2022;**134**:127-136

[25] Majhi SM, Mirzaei A, Kim HW, Kim SS, Kim TW. Recent advances in energy-saving chemiresistive gas sensors: A review. Nano Energy. 2021;**79**:105369

[26] Saxena P, Shukla P. A review on gas sensor technology and its applications. In: Rao VV, Kumaraswamy A, Kalra S, Saxena A, editors. Computational and Experimental Methods in Mechanical Engineering. Singapore: Springer Singapore; 2022. pp. 165-175

[27] Ren F, Pearton SJ, Pearton S, Ren F. Semiconductor Device-Based Sensors for Gas, Chemical, and Biomedical Applications. Raton, FL: CRC Press Boca; 2011

[28] Sunil M, Shweta J. Metal-oxide semiconductors for carbon monoxide (CO) gas sensing: A review. Applied Materials Today. 2020;**18**:100483

[29] Jacobson MZ. Short-term effects of controlling fossil-fuel soot, biofuel

soot and gases, and methane on climate, Arctic ice, and air pollution health. Journal of Geophysical Research: Atmospheres. 2010;**115**:D14209. DOI: 10.1029/2009JD013795

[30] Babariya B, Raval D, Gupta SK, Gajjar P. Selective and sensitive toxic gas-sensing mechanism in a 2D Janus MoSSe monolayer. Physical Chemistry Chemical Physics. 2022;**24**(25):15292-15304

[31] Yunusa Z, Hamidon MN, Kaiser A, Awang Z. Gas sensors: A review. Sens Transducers. 2014;**168**(4):61-75

[32] Solórzano A, Eichmann J, Fernández L, Ziems B, Jiménez-Soto JM, Marco S, et al. Early fire detection based on gas sensor arrays: Multivariate calibration and validation. Sensors and Actuators B: Chemical. 2022;**352**: 130961

[33] Mishra V, Rashmi, Sukriti. Optical Gas Sensors. London, UK: IntechOpen; 2023. Available from: https://www.intechopen.com/chapters/85155

[34] Paliwal A, Sharma A, Tomar M, Gupta V. Carbon monoxide (CO) optical gas sensor based on ZnO thin films. Sensors and Actuators B: Chemical. 2017;**250**:679-685

[35] Sridhar AS, Chen X, Glossmann T, Yang Z, Xu Y, Lai W, et al. Single-frequency impedance studies on an ionic liquid-based miniaturized electrochemical sensor toward continuous low-temperature CO_2 monitoring. ACS Sensors. 2023;**8**(1):197-206

[36] Miralaei M, Salari S, Kameli P, Goodarzi MT, Ranjbar M. Electrical and hydrogen gas sensing properties of Co1-xZnxFe_2O_4 nanoparticles; effect of the sputtered palladium thin layer. International Journal of Hydrogen Energy. 2023. Available from: https://www.sciencedirect.com/science/article/pii/S0360319923007619

[37] Paghi A, Mariani S, Barillaro G. 1D and 2D field effect transistors in gas sensing: A comprehensive review. Small. 2023:2206100. Available from: https://onlinelibrary.wiley.com/doi/full/10.1002/smll.202206100

[38] Mirzaei A, Lee J-H, Majhi SM, Weber M, Bechelany M, Kim HW, et al. Resistive gas sensors based on metal-oxide nanowires. Journal of Applied Physics. 2019;**126**(24):241102

[39] Pourteimoor S, Haratizadeh H. Performance of a fabricated nanocomposite-based capacitive gas sensor at room temperature. Journal of Materials Science: Materials in Electronics. 2017;**28**:18529-18534

[40] Wang Z, Zhu L, Wang J, Zhuang R, Mu P, Wang J, et al. Advances in functional guest materials for resistive gas sensors. RSC advances. 2022;**12**(38):24614-24632

[41] John RAB, Ruban KA. A review on resistive-based gas sensors for the detection of volatile organic compounds using metal-oxide nanostructures. Inorganic Chemistry Communications. 2021;**133**:108893

[42] Yang Z, Jiang L, Wang J, Liu F, He J, Liu A, et al. Flexible resistive NO_2 gas sensor of three-dimensional crumpled MXene Ti_3C_2Tx/ZnO spheres for room temperature application. Sensors and Actuators B: Chemical. 2021;**326**:128828

[43] Neri, Giovanni, Donato N. Resistive gas sensors. Wiley Encyclopedia of Electrical and Electronics Engineering. 1999:1-12

[44] Wang C, Yin L, Zhang L, Xiang D, Gao R. Metal oxide gas sensors: Sensitivity and influencing factors. Sensors (Basel). 2010;**10**(3):2088-2106

[45] Zhang C, Xu K, Liu K, Xu J, Zheng Z. Metal oxide resistive sensors for carbon dioxide detection. Coordination Chemistry Reviews. 1 Dec 2022;**472**:214758

[46] Kim H-J, Lee J-H. Highly sensitive and selective gas sensors using p-type oxide semiconductors: Overview. Sensors and Actuators B: Chemical. 2014;**192**:607-627

[47] Miller DR, Akbar SA, Morris PA. Nanoscale metal oxide-based heterojunctions for gas sensing: A review. Sensors and Actuators B: Chemical. 2014;**204**:250-272

[48] Dobrokhotov V, Larin A, Sowell D. Vapor trace recognition using a single nonspecific chemiresistor. Sensors (Basel). 2013;**13**(7):9016-9028

[49] Righettoni M, Amann A, Pratsinis SE. Breath analysis by nanostructured metal oxides as chemo-resistive gas sensors. Materials Today. 2015;**18**(3):163-171

[50] Jian Y, Hu W, Zhao Z, Cheng P, Haick H, Yao M, et al. Gas sensors based on chemi-resistive hybrid functional nanomaterials. Nano-Micro Letters. 2020;**12**:1-43

[51] Dadkhah M, Tulliani JM. Green synthesis of metal oxides semiconductors for gas sensing applications. Sensors (Basel). 2022;**22**(13):4669

[52] Siebert L, Wolff N, Ababii N, Terasa M-I, Lupan O, Vahl A, et al. Facile fabrication of semiconducting oxide nanostructures by direct ink writing of readily available metal microparticles and their application as low power acetone gas sensors. Nano Energy. 2020;**70**:104420

[53] Zhou S, Gong S, Zhong S, Pan W, Ying W, editors. Region selection model with saliency constraint for fine-grained recognition. In: Neural Information Processing. Cham: Springer International Publishing; 2019

[54] Lee, Eunji, Yoon YS, Kim D-J. Two-dimensional transition metal dichalcogenides and metal oxide hybrids for gas sensing. ACS Sensors. 2018;**3**(10):2045-2060

[55] Bhati VS, Kumar M, Banerjee R. Gas sensing performance of 2D nanomaterials/metal oxide nanocomposites: A review. Journal of Materials Chemistry C. 2021;**9**(28):8776-8808

[56] Tang Y, Zhao Y, Liu H. Room-temperature semiconductor gas sensors: Challenges and opportunities. ACS Sensors. 2022;**7**(12):3582-3597

[57] Pandit NA, Ahmad T. Tin oxide based hybrid nanostructures for efficient gas sensing. Molecules. 2022;**27**(20):7038

[58] Parashar M, Shukla VK, Singh R. Metal oxides nanoparticles via sol-gel method: A review on synthesis, characterization and applications. Journal of Materials Science: Materials in Electronics. 2020;**31**(5):3729-3749

[59] Li ZJ, Li H, Wu ZL, Wang MK, Luo JT, Torun HD, et al. Advances in designs and mechanisms of semiconducting metal oxide nanostructures for high-precision gas sensors operated at room temperature. Materials Horizons. 2019;**6**(3):470-506

[60] Li T, Yin W, Gao SW, Sun YN, Xu PL, Wu SH, et al. The combination of two-dimensional nanomaterials with metal oxide nanoparticles for gas sensors: A review. Nanomaterials. 2022;**12**(6):982

[61] Panday M, Upadhyay GK, Purohit L. Sb incorporated SnO_2 nanostructured thin films for CO_2 gas sensing and humidity sensing applications. Journal of Alloys and Compounds. 2022;**904**:164053

[62] Basyooni MA, Shaban M, El Sayed AM. Enhanced gas sensing properties of spin-coated Na-doped ZnO nanostructured films. Scientific Reports. 2017;7(1):1-12

[63] Sonker RK, Sabhajeet S, Yadav B. TiO_2–PANI nanocomposite thin film prepared by spin coating technique working as room temperature CO_2 gas sensing. Journal of Materials Science: Materials in Electronics. 2016;**27**:11726-11732

[64] Choudhary K, Saini R, Upadhyay GK, Purohit L. Sustainable behavior of cauliflower like morphology of Y-doped ZnO: CdO nanocomposite thin films for CO_2 gas sensing application at low operating temperature. Journal of Alloys and Compounds. 2021;**879**:160479

[65] Chen Y, Wang D, Qin H, Zhang H, Zhang Z, Zhou G, et al. CO_2 sensing properties and mechanism of $PrFeO_3$ and $NdFeO_3$ thick film sensor. Journal of Rare Earths. 2019;**37**(1):80-87

[66] Singh A, Singh A, Singh S, Tandon P, Yadav B. Preparation and characterization of nanocrystalline nickel ferrite thin films for development of a gas sensor at room temperature. Journal of Materials Science: Materials in Electronics. 2016;**27**:8047-8054

[67] Susanti D, Diputra AGP, Tananta L, Purwaningsih H, Kusuma GE, Wang C, et al. WO3 nanomaterials synthesized via a sol-gel method and calcination for use as a CO gas sensor. Frontiers of Chemical Science and Engineering. 2014;**8**:179-187

[68] Zhang T, Liu L, Qi Q, Li S, Lu G. Development of microstructure In/Pd-doped SnO_2 sensor for low-level CO detection. Sensors and Actuators B: Chemical. 2009;**139**(2):287-291

[69] Mohammadi M, Fray D. Nanostructured TiO_2–CeO_2 mixed oxides by an aqueous sol–gel process: Effect of Ce: Ti molar ratio on physical and sensing properties. Sensors and Actuators B: Chemical. 2010;**150**(2):631-640

[70] Mohammadi M, Fray D. Synthesis and characterisation of nanosized TiO_2–ZrO_2 binary system prepared by an aqueous sol–gel process: Physical and sensing properties. Sensors and Actuators B: Chemical. 2011;**155**(2):568-576

[71] Singh N, Gupta RK, Lee PS. Gold-nanoparticle-functionalized In_2O_3 nanowires as CO gas sensors with a significant enhancement in response. ACS applied materials & interfaces. 2011;**3**(7):2246-2252

[72] Hsu K-C, Fang T-H, Hsiao Y-J, Wu P-C. Response and characteristics of TiO_2/perovskite heterojunctions for CO gas sensors. Journal of Alloys and Compounds. 2019;**794**:576-584

[73] Hübner M, Simion C, Haensch A, Barsan N, Weimar U. CO sensing mechanism with WO3 based gas sensors. Sensors and Actuators B: Chemical. 2010;**151**(1):103-106

[74] Yang W-D, Chang Y-H, Huang C-C, Chen Y-C. Microstructure and characteristics of thin-film La 0.8 Sr 0.2 Co 0.5 Ni 0.5 O3/ZnO: Al Heterocontact CO sensors prepared by RF magnetron sputtering. Journal of Electronic Materials. 2009;**38**:460-467

[75] Hsu K-C, Fang T-H, Chen S-H, Kuo E-Y. Gas sensitivity and sensing mechanism studies on ZnO/La0. 8Sr0. 2Co0. 5Ni0. 5O3 heterojunction

structure. Ceramics International. 2019;**45**(7):8744-8749

[76] Chen Y, Chang Y-H, Chen G-J, Chai Y-L, Ray D. The sensing properties of heterojunction SnO2/La0. 8Sr0. 2Co0. 5Ni0. 5O3 thin-film CO sensor. Sensors and Actuators B: Chemical. 2003;**96**(1-2):82-87

[77] Yan WJ, Hu M, Zeng P, Ma SY, Li MD. Room temperature NO_2-sensing properties of WO3 nanoparticles/porous silicon. Applied Surface Science. 2014;**292**:551-555

[78] Gawali SR, Patil VL, Deonikar VG, Patil SS, Patil DR, Patil PS, et al. Ce doped NiO nanoparticles as selective NO2 gas sensor. Journal of Physics and Chemistry of Solids. 2018;**114**:28-35

[79] Giancaterini L, Cantalini C, Cittadini M, Sturaro M, Guglielmi M, Martucci A, et al. Au and Pt nanoparticles effects on the optical and electrical gas sensing properties of sol–gel-based ZnO thin-film sensors. IEEE Sensors Journal. 2014;**15**(2):1068-1076

[80] Haiduk YS, Khort A, Lapchuk N, Savitsky A. Study of WO3–In_2O_3 nanocomposites for highly sensitive CO and NO_2 gas sensors. Journal of Solid State Chemistry. 2019;**273**:25-31

[81] Kaur J, Kumar R, Bhatnagar M. Effect of indium-doped SnO_2 nanoparticles on NO_2 gas sensing properties. Sensors and Actuators B: Chemical. 2007;**126**(2):478-484

[82] Kumar R, Kumari R, Singh VN. SnO_2-based NO_2 gas sensor with outstanding sensing performance at room temperature. Micromachines. 2023;**14**(4):728

[83] Jie XQ, Zeng DW, Zhang J, Xu K, Wu JJ, Zhu BK, et al. Graphene-wrapped WO3 nanospheres with room-temperature NO_2 sensing induced by interface charge transfer. Sensors and Actuators B: Chemical. 2015;**220**:201-209

[84] Abruzzi RC, Pires MJR, Dedavid BA, Galli CF. Application of SnO_2 nanoparticles and zeolites in coal mine methane sensors. Materials Research. 2019;**22**

[85] Bhattacharyya P, Biswas S, Sengupta A, Maji T, Saha H. Palladium surface modification of Nanocrystalline sol-gel derived zinc oxide thin films and its effect on methane sensing. Sensors & Transducers. 2009;**110**(11):38

[86] Anchal BBN, Singh P, Pyare R. A Nano-wrinkled Zn0. 92Fe0. 08O thin film developed using a high-RPM electro-spin patterning technique via sol-gel route for methane sensing. ChemistrySelect. 2018;**3**(42):11881-11889

[87] Giang HT, Duy HT, Ngan PQ, Thai GH, Toan NN. Hydrocarbon gas sensing of nano-crystalline perovskite oxides $LnFeO_3$ (ln= La, Nd and Sm). Sensors and Actuators B: Chemical. 2011;**158**(1):246-251

[88] Liu D, Lin L, Chen Q, Zhou H, Wu J. Low power consumption gas sensor created from silicon nanowires/TiO_2 core–shell heterojunctions. ACS sensors. 2017;**2**(10):1491-1497

[89] Chesler P, Hornoiu C, Anastasescu M, Calderon-Moreno JM, Gheorghe M, Gartner M. Cobalt-and copper-based Chemiresistors for low concentration methane detection, a comparison study. Gels. 2022;**8**(11):721